Lecture Notes on Data Engineering and Communications Technologies

Volume 94

Series Editor
Fatos Xhafa, Technical University of Catalonia, Barcelona, Spain

The aim of the book series is to present cutting edge engineering approaches to data technologies and communications. It will publish latest advances on the engineering task of building and deploying distributed, scalable and reliable data infrastructures and communication systems.

The series will have a prominent applied focus on data technologies and communications with aim to promote the bridging from fundamental research on data science and networking to data engineering and communications that lead to industry products, business knowledge and standardisation.

Indexed by SCOPUS, INSPEC, EI Compendex.

All books published in the series are submitted for consideration in Web of Science.

More information about this series at https://link.springer.com/bookseries/15362

Isaac Woungang • Sanjay Kumar Dhurandher
Editors

4th International Conference on Wireless, Intelligent and Distributed Environment for Communication

WIDECOM 2021

Springer

Editors
Isaac Woungang
Department of Computer Science
Ryerson University
Toronto, ON, Canada

Sanjay Kumar Dhurandher
Department of Information Technology
Netaji Subhas University of Technology
New Delhi, Delhi, India

ISSN 2367-4512 ISSN 2367-4520 (electronic)
Lecture Notes on Data Engineering and Communications Technologies
ISBN 978-3-030-89775-8 ISBN 978-3-030-89776-5 (eBook)
https://doi.org/10.1007/978-3-030-89776-5

This Springer imprint is published by the registered company Springer Nature Switzerland AG
The registered company address is: Gewerbestrasse 11, 6330 Cham, Switzerland

Preface

The last decade has witnessed tremendous advances in computing and networking technologies, with the appearance of new paradigms such as Internet of Things (IoT) and cloud computing, which have led to advances in wireless and intelligent systems for communications. Undoubtedly, these technological advances help improve many facets of human lives, for instance, through better healthcare delivery, faster and more reliable communications, significant gains in productivity, to name a few. At the same time, the associated increasing demand for a flexible and cheap infrastructure for collecting and monitoring real-world data nearly everywhere and every time, coupled with the above-mentioned integration of wireless mobile systems and network computing, raises new challenges with respect to the dependability of integrated applications and the intelligence-driven security threats against the platforms supporting these applications. The WIDECOM conference is a conference series that provides a venue for researchers and practitioners to present, learn, and discuss recent advances in new dependability paradigms, design, and performance of dependable network computing and mobile systems, as well as issues related to the security of these systems.

Toronto, ON, Canada — Isaac Woungang
New Delhi, Delhi, India — Sanjay Kumar Dhurandher

Contents

1 **Adaptive System for Object Detection and Picking Based on Efficient Convolutional Neural Networks** ... 1
Hwang-Cheng Wang, Wei-Zhi Chen, Yan-Long Huang, and Jia-Jun Zhuang

2 **Cybersecurity Data Science: Concepts, Algorithms, and Applications** ... 21
Wei Lu

3 **Comparison of Task Scheduling Algorithms for Traffic Surveillance Application Using Fog Computing** ... 31
Mluleki Sinqadu, Zelalem Sintayehu Shibeshi, and Khuram Khalid

4 **Modifying the SMOTE and Safe-Level SMOTE Oversampling Method to Improve Performance** ... 47
Kgaugelo Moses Dolo and Ernest Mnkandla

5 **Performance Evaluation of Fuzzy-Based Routing Protocols for Opportunistic Networks** ... 61
Khuram Khalid, Isaac Woungang, Sanjay K. Dhurandher, and Jagdeep Singh

6 **ACIDS: A Secure Smart City Framework and Threat Model** ... 79
Soomaiya Hamid and Narmeen Zakaria Bawany

7 **Volunteer Drone: Search and Rescue of the Industrial Building Collapsed Worker** ... 99
A. K. M. Islam, Dalia Hanna, and Alexander Ferworn

8 **Optimizing the Key-Pair Generation Phase of McEliece Cryptosystem** ... 111
Michela Ceria, Alessandro De Piccoli, Martino Tiziani, and Andrea Visconti

9 Dual Parameter Ranking Based Resource Allocation for PD-SCMA Cognitive Radio Networks 123
Simon Chege and Tom Walingo

Index 139

Chapter 1
Adaptive System for Object Detection and Picking Based on Efficient Convolutional Neural Networks

Hwang-Cheng Wang, Wei-Zhi Chen, Yan-Long Huang, and Jia-Jun Zhuang

1.1 Introduction

This work aims to improve environmental cleanliness by constructing an adaptive object detection and recognition system. The system is composed of the following main components:

Robotic arm control: The control mechanism directs a multi-axis robotic arm to precisely pick up garbage and moves it into a bin at the back of the robot.

Positioning and path planning: In the outdoors, GPS is used to obtain position information and guides the robot. In indoor areas or areas where satellite signals cannot be received, paths are planned through machine motion trajectories and perimeter sensors.

Automatic patrol: After identifying the location of garbage, the robot automatically navigates to the target.

Database management: The database increases data management efficiency so that relevant parties can easily query the operation data. Moreover, through big data analysis, the main distribution area of garbage can be used to determine the best movement path.

Remote monitoring, control, and security: Through the technology of Internet of Things, it is possible to control and monitor the operation of the robot. Under special conditions, such as rain, large garbage that cannot be transported, or threat to the robot, suitable information will be sent to the operator.

The overall system architecture is shown in Fig. 1.1.

The rest of the chapter is organized as follows. In Sect. 1.2, the motivation for the work is described. In Sect. 1.3, the design methodology is presented. In Sect. 1.4, the

H.-C. Wang (✉) · W.-Z. Chen · Y.-L. Huang · J.-J. Zhuang
National Ilan University, Yilan, Taiwan
e-mail: hcwang@niu.edu.tw; b0742006@ems.niu.edu.tw; b0742017@ems.niu.edu.tw

I. Woungang, S. K. Dhurandher, *4th International Conference on Wireless, Intelligent and Distributed Environment for Communication*, Lecture Notes on Data Engineering and Communications Technologies 94, https://doi.org/10.1007/978-3-030-89776-5_1

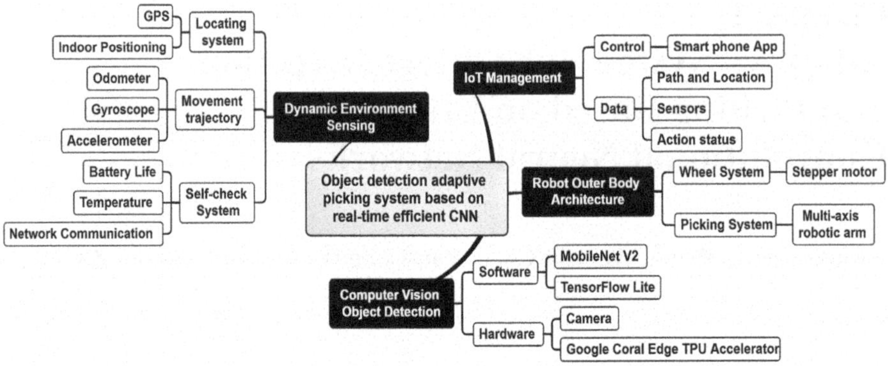

Fig. 1.1 System architecture

hardware and software implementation are discussed. In Sect. 1.5, the results and optimization of the proposed scheme are presented. Finally, Section 1.6 concludes the chapter and points to future work.

1.2 Motivation

This work was inspired by the innovative combination of technological advances to address a social issue. As the renowned economist John Galbraith eloquently put it, technology means the systematic application of scientific or other organized knowledge to practical tasks. There has been a flurry of new development in the areas of machine learning, wireless communication, and Internet of Things. The robot described in the chapter exploits machine learning to quickly identify common types of garbage that can be easily spotted on the street. It was trained using an efficient convolutional neural network. A camera mounted on the robot picks up images of objects on the street, and the robot will move toward an object if it is determined to be garbage. Various sensors embedded in the robot allow it to sense the surrounding area. Wireless communication is used to send signals among the components of the robot using a lightweight messaging protocol. The agile movement of the robot is enabled by inverse kinematic analysis. A robot operating system facilitates system integration which involves many hardware and software components. A number of measures have been taken to improve the dependability and robustness of the robot. Power consumption is an important design parameter. Innovative approaches have been adopted to reduce the power consumption to avoid the quick depletion of the battery that drives the robot. Finally, the design and implementation of the robot aim to improve environmental cleanness, which is vital to public hygiene.

1.3 Design Methodology

In order to reduce the cost of maintaining a clean environment and improve efficiency, we rely on the rapid development of image recognition and sorting technology in recent years and realize waste picking automation through image recognition and robotic arm. We exploit the classification and identification of garbage and integrate the robotic arm and wheel system to move swiftly to the target location. Through machine learning and big data analysis, we link computer vision and robot action to achieve intelligent autonomous garbage picking. Compared to the general sweeping robot, our design can handle a greater variety of garbage and can adapt to diverse terrains and environments.

Algorithm This work aims to recognize garbage, and we use a wide-angle lens as the main component for environmental image collection. In 2017, Google proposed MobilenetV1, which enabled CNNs to achieve higher accuracy with fewer computational resources through depthwise separable convolution. Based on MobilenetV1, Google proposed MobilenetV2 in 2018, which made CNNs more lightweight and achieved higher accuracy at the same time. Because of these, we adopted MobileNetV2 as the primary method to identify garbage.

This method can segment the image contents and mark them as individual objects and select and track specific objects, and finally assign appropriate labels to the computation results. Of course, the naming and rules of the tags can be decided and designed by the developer.

TPU In order to improve the performance of image recognition, we used Edge TPU Accelerator (Fig. 1.2) from Google as the processing core for real-time detection, together with the lightweight artificial intelligence framework TensorFlow Lite. TensorFlow Lite, a lightweight solution designed for mobile devices and embedded systems, can convert almost all the models trained in TensorFlow into TensorFlow Lite files (.tflite). The hardware and software eased the task of image detection.

Machine Learning and Visualization The machine learning was carried out on a Dell EMC PowerEdge R740 rackmount server with two Intel Xeon Gold 5218 (64-core processors) and two NVIDIA Quadro RTX™ 6000s to allow a large number of deep learning tasks to be performed more efficiently. The training also established checkpoints, which can be used to create a new set of tasks. The training also created checkpoints and visualized the training status through the TensorBoard streaming webpage (Fig. 1.3). The test results of image recognition are shown in Fig. 1.4.

Distance Sensor Optical sensors of reflective model and ultrasonic sensors were used to measure the distance.

We used the Adafruit VL53L0X time-of-flight distance sensor and HC-SR04 ultrasonic distance sensor to perform distance measurements and obtained the results in Table 1.1.

The optical distance sensor is more accurate than the ultrasonic counterpart and was used to measure the distance between an object and the lens. On the other

Fig. 1.2 Edge TPU accelerator

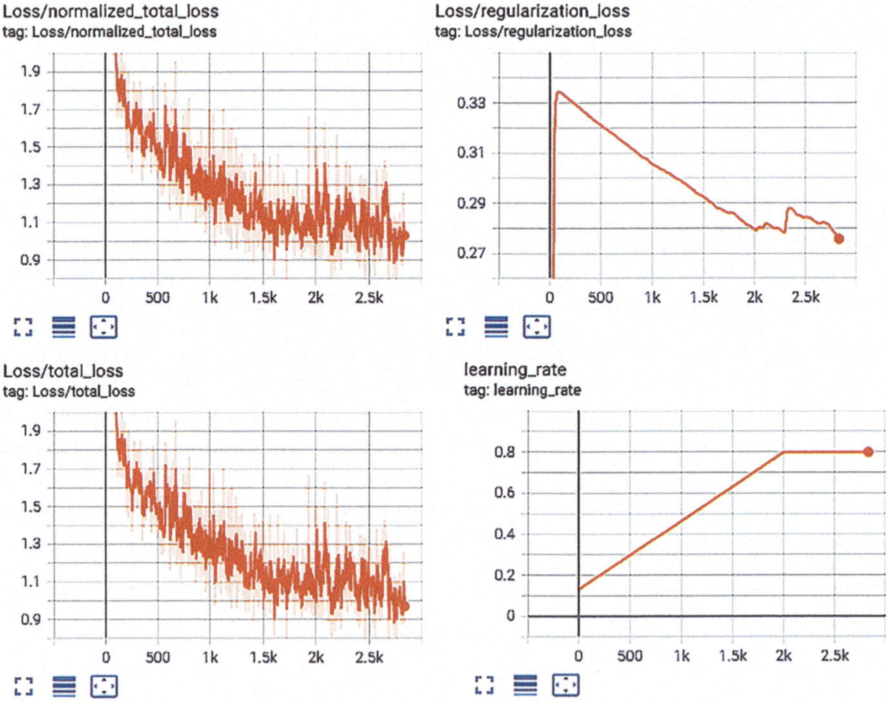

Fig. 1.3 Graphs showing the loss function and learning rate

Fig. 1.4 Result of image detection

Table 1.1 Comparison of optical ranging and ultrasonic ranging modules

Module type	Optical (reflective model)	Ultrasonic
Representative module	VL53L0X	HC-SR04
Detectable target	Detection is affected by target materials/colors	Detection is unaffected by target materials/colors
Detecting distance	1000 mm max	4 m max
Accuracy	High	Low
Response speed	Fast	Slow
Dust/water	Affected	Unaffected
Measuring range	Small	Large

hand, the ultrasonic sensor has a broader coverage and was used for robot proximity sensing.

1.4 Hardware and Software Implementation

1.4.1 Integrated Development Environment

The Raspberry Pi development board was used to accommodate the components of the system. Image recognition constituted the core of the machine vision block.

The rest of the control, action block, and so on also needed to be incorporated into the overall architecture. A ROS development environment was built on top of the Ubuntu 20.04 system. The environment allowed the development and testing of individual functional blocks. The integration of the functional blocks is also easier and faster under ROS. In analogy to the human body, the system possessed the following functions:

Eye for image recognition: classification and recognition of targets through computer vision and machine learning.

Brain for environment integration: as the main control board connecting the hub of each block and the environment system mounted on it.

Hand for object picking: picking up the target by the robotic arm.

Feet for system mobility: moving along the path on wheels.

Ear for remote monitoring: receiving control and returning data through the Internet of Things technology.

Nerve for transmission of control signals: controlling various sensors, power management, etc.

The key components are described in the following.

Raspberry Pi The Raspberry Pi 4 Model B development platform is equipped with a quad-core 1.5 GHz ARM Cortex-A72 Broadcom BCM2711 processor, which provides excellent system control. It is fitted with HAT network power supply (PoE). Besides, it has 802.11 ac dual-band WiFi and Bluetooth 5.0 for wireless communication. In addition, it is lightweight and much cheaper than ordinary computers. It also supports many common I/Os.

Ubuntu In order to recognize objects of interest, the machine learning model needs to be trained. The pre-processing requires a development environment with different kits and compilers depending on the characteristics of the ported system. The environment for the training of the image recognition model is Ubuntu 20.04 operating system, which is the most widely used version of Linux at present. The Kernel Image generates smaller files according to different requirements, which is very beneficial for embedded products with hardware imitations.

ROS ROS is short for Robot Operating System developed from 2010 to now. Ubuntu 20.04 uses the 13th release of ROS Noetic Ninjemys, which is mainly responsible for connecting various components inside the robot for easy and fast connection. ROS supports many programming languages. The internal library tools of ROS also make maintenance easier. Overall, ROS is very suitable for large-scale real-time systems.

1.4.2 Data Preprocessing

The dataset contains five labels corresponding to bottles, masks, trash, glass, and cans in this work. Herein trash specifically refers to objects that are kneaded

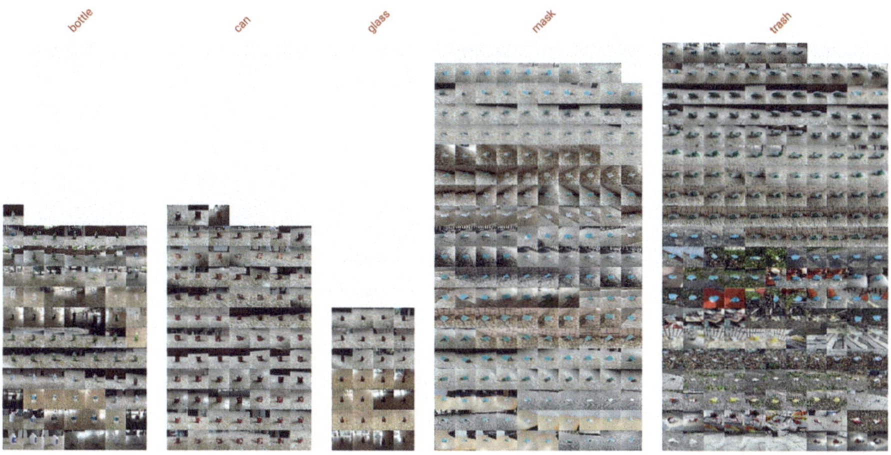

Fig. 1.5 Images used for training

into balls or objects with irregular shapes, such as plastic bags, crumpled paper, and paper balls. These are typically considered to be garbage and are frequently encountered in the surroundings. Figure 1.5 demonstrates part of the dataset.

We took about 3000 pictures of these objects by smartphone. Since the architecture of MobileNetV2 contains a fully convolution layer [1], it is reasonable to resize these images to 640×360 rather than a square. It helps retain more information from images. Moreover, the smaller size of images requires less computing time for training.

To ensure that the training and testing sets have approximately the same percentage of each target class sample in the dataset, we applied stratified sampling to split the data into a ratio of 8 to 2, which are the proportion of the training and testing sets. To avoid overfitting and to enhance the features of images of interest, we also made good use of OpenCV, which helped us create an augmented dataset. The augmentation consists of random Gaussian blur of between 0 and 0.75 pixels, salt and pepper noise applied to 5% of the pixels, and random exposure adjustment between -17% and $+17\%$ [2–4].

A desirable feature of the model is its scalability. If one needs to increase the recognition targets, one only needs to expand the dataset and retrain the model.

1.4.3 Remote Monitoring

Mobile Device Applications Nowadays, smartphones are ubiquitous and have excellent connection and control capabilities. With the appropriate APP installed, they can support the remote operation of the database and robot. Android Studio was used for the development of the APP. Users employed the APP to control the robot, and the operation relied on the establishment of MQTT connections. MQTT is a simple publish/subscribe messaging protocol designed for devices with low hardware performance and poor network conditions.

In addition, the use of Fragment [5] in multi-screen switching reduces the resources and time expended on object reconstruction. It brings the objects to the interface of each navigation through the helper classes of ViewModel, which makes the overall APP smoother and more stable.

Database Data is accumulated into a large amount over a long period. Without a database, access and modification would be cumbersome. The use of a database brings many benefits, such as unified data format, easy access, and simplified data operations (add, modify, delete, query, etc.). A variety of software is available for this purpose. One just needs to learn the procedure to enhance the development efficiency greatly. However, to avoid the heavy network load caused by continuous data upload, a lightweight version of MySQL, SQLite, was installed on the main control board as a tentative data storage. Data stored on SQLite is synchronized to the main MySQL database periodically. Since the two are compatible, SQLite data can be sent to MySQL directly [6, 7].

1.4.4 Object-Picking Subsystem

The object-picking subsystem comprises the jaw, arm, and control (Fig. 1.6).

Jaw The three essential requirements on the jaws are the exact clamping stability of the object, sufficient opening and closing range, and self-adjustment of excessive

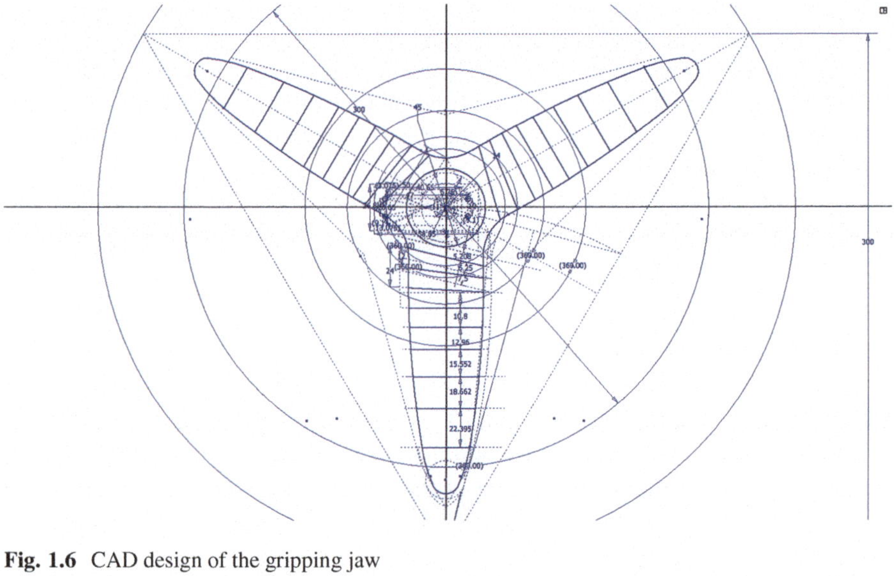

Fig. 1.6 CAD design of the gripping jaw

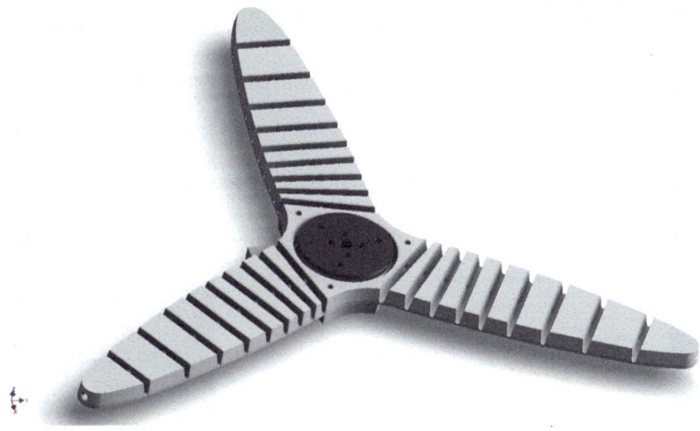

Fig. 1.7 Gripping jaw in final form

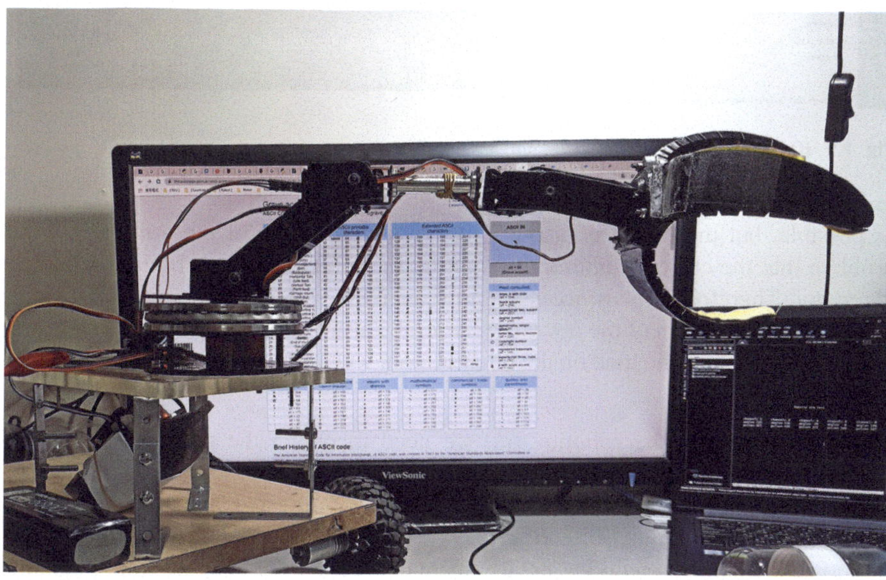

Fig. 1.8 Five-axis robotic arm

force. The jaws must clamp objects stably. Commercially available two-finger jaws can clamp objects firmly, but the acceptable object size is limited. On the other hand, three-finger jaws can grip objects like 1-liter bottles. However, because of the simple magnetic force used to close the jaws, they can only pick up lighter items. To make up for the limited clamping range of the commercially available two-finger jaws and the clamping force of the three-finger jaws, we have constructed a three-finger jaw, as shown in Fig. 1.7, by referring to the materials in Thingiverse, a sharing design platform for 3D printing. The jaw can be retracted by a single servo motor.

Arm The robotic arm, as shown in Fig. 1.8, was assembled from the support bracket and the jaw described above. The skeleton structure has been made as light

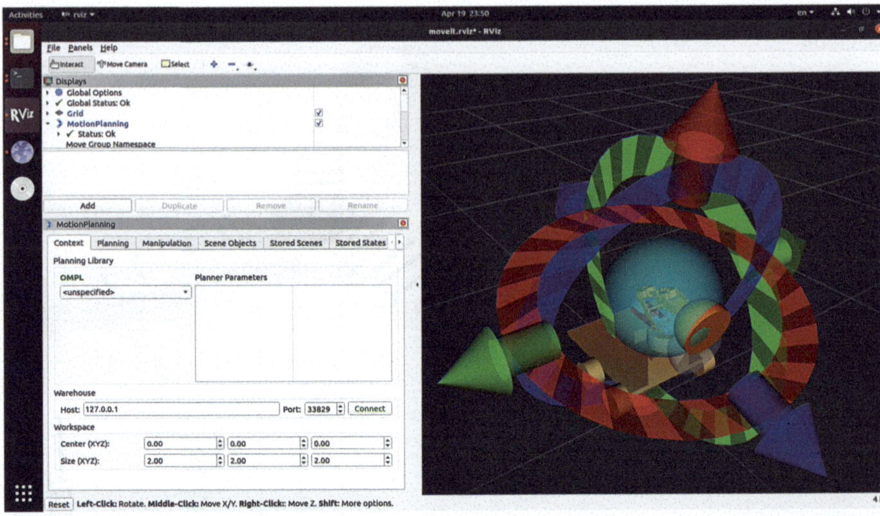

Fig. 1.9 Rviz arm posture motion planning

as possible, but the motor is heavy. The original structure of the arm caused the problem that the current running through the motor exceeded the limit and burned the motor. Therefore, it was necessary to reduce the weight of the arm and change the degrees of freedom of the jaw wrist. After the arm's center of gravity was shifted to the base to reduce the load on the other motors when lifting, the arm became more stable.

Control To control the robotic arm, we need to obtain the path angle of each arm posture first and then obtain the rotation amplitude of each motor by inverse kinematic analysis. Inverse kinematic analysis is complicated. We used MoveIt [8, 9], a ROS programming suite (Fig. 1.9) with kinematic forward and backward solutions, collision detection, environmental recognition, and motion planning algorithms, to accomplish motion planning of the arm posture. We then programmed the calculated values to the robotic arm to achieve the ideal arm motion.

1.4.5 MoveIt Robot Control Steps

Building the Robot URDF Model File The robot form was drawn to scale using Autodesk Fusion 360 3D CAD software. Then the open-source Python script fusion2urdf converted the drawn robot body drawing into a URDF (Unified Robot Description Format) file, which included linkage, joint names, kinematic parameters, kinetic parameters, visualization model, collision detection model, etc.

Building Robotic ROS Drive A ROS driver was written to provide an action server to receive the results planned by MoveIt, send them to the robot, and provide feedback during execution.

Generating MoveIt Configuration Files We used the setup assistant interface of MoveIt to configure the simulation environment. The effect of motion planning could be viewed in the simulation environment Rviz.

Calibrating the Camera This step set the position of the camera relative to the robot itself in the launch file.

Modifying MoveIt Configuration File and Launch File Parameters used in the default profile generated by MoveIt for the virtual robot were modified to send signals that control the real robot.

1.4.6 Electrical Current Sensing and Adaptive Control

Servo motors were used for arm and jaw control. A servo motor can control the angle precisely but cannot know the pressure applied. If the motor gets jammed, it will be overloaded, which will cause the speed of the motor to drop, the current to increase, the temperature to rise, and the winding coils to overheat. The force applied by the jaw during clamping and the weight of the robotic arm during lifting were tested with the INA219 electric current measurement module. The module measured the current of the servo motor to know its status [10]. When the current of the servo motor gradually rises from 0.01 A to 1.7 A, slight rotation of the motor angle will cause the current to jump to more than 2 A abruptly. At this time, the motor must be stopped. From the measurement result, we can tell whether the clamping jaw is in a tightly clamped state. This adaptive design avoids damage to the motor and mechanical structure to make the system more dependable.

1.4.7 Mobile Subsystem

Body Movement Design The off-road tires have enough grip to adapt to various ground conditions. But adaptation to rugged terrain relies on the suspension system for each tire to move smoothly. It was envisioned that the NASA moon vehicle's rocker-bogie system for design and assembly would allow large structures to climb stairs. However, such a design might hinder the freedom of mechanical arm extension, so the wheeled system design with a low chassis was used (Fig. 1.10).

Path Tracking Outdoor positioning can be achieved directly through GPS, while indoor space cannot receive satellite signals. Simultaneous localization and mapping (SLAM) technology, which starts from an unknown location in an unknown environment, locates its own position and attitude by repeatedly observing map

Fig. 1.10 Design of the wheeled system with low chassis

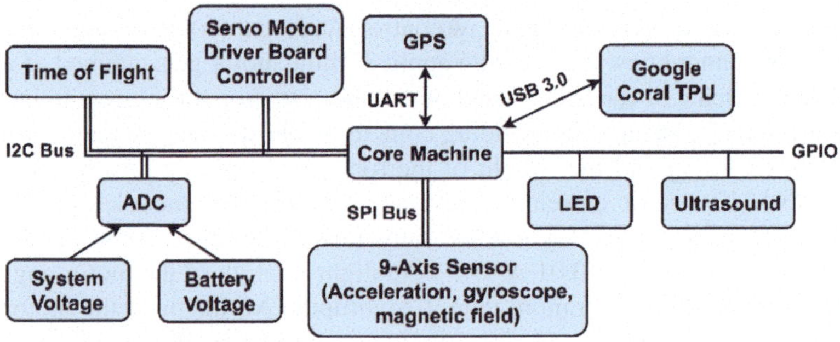

Fig. 1.11 Logic circuit architecture

features during motion and then constructs a map incrementally based on its own position, thereby achieving the purpose of simultaneous localization and map construction [11–13]. This technique was realized by ROS mentioned above.

Circuit Design To communicate with the many sensors and function modules, we have used the various interfaces supported by Raspberry Pi 4. The main control board is only responsible for communication with the interfaces, while communication with the sensors and calculations are performed on the individual modules. The arrangement can reduce the load on the main control board and simplify the circuit wiring, which is very convenient for design and debugging. The logic circuit architecture is shown in Fig. 1.11.

Power Supply Subsystem The power was supplied by lithium-ion polymer batteries and divided into high-load and low-load parts. The battery protection module

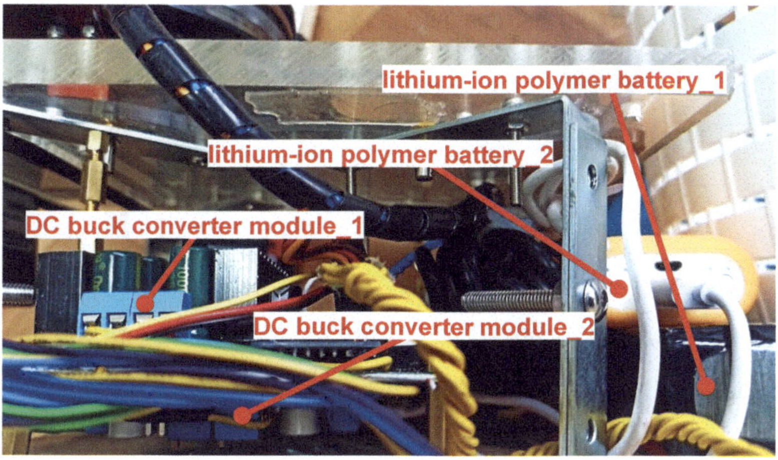

Fig. 1.12 Dual power supply system in the circuit board design

and DC step-down module were used to prevent the two parts from interfering with each other due to different power requirements [14].

The main power supply of the lithium battery output was divided into two isolated power sources. The load required by the robot can be separated into two categories, one with low and stable power consumption by the computing IC components and the other with high current and considerable variation by the motors. It was found that a slight change during transient in the motor current would have a significant impact on the main control board, causing it to reset. To solve the problem, we changed the design to two isolated power supply paths. Doing so greatly improved the stability of the system. The test circuit board is shown in Fig. 1.12.

1.5 Results and Optimization

1.5.1 Image Recognition

After several rounds of training and parameter tuning, the optimal results were finally achieved. The image recognition results of our model are shown in Fig. 1.13. Overall, the model performs well. However, the precision in recognizing bottles and glass is lower, as reflected in Table 1.2 and Fig. 1.14, which implies that the model might sometimes fail while detecting bottles and glass jars on the street. There are possibly two reasons for the low recognition rate of bottles and glass:

1. The training samples are homogeneous and large in number, leading to overfitting [15].

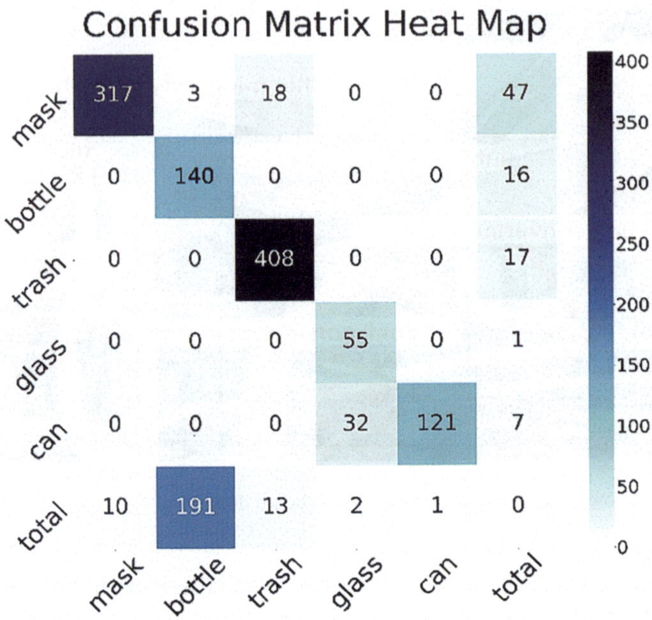

Fig. 1.13 Confusion matrix. The accuracy of the model is 0.8807

Table 1.2 Statistics of the evaluation

Category	Precision_@0.5IOU	Recall_@0.5IOU	F1_score
Mask	0.969419	0.823377	0.890449
Bottle	0.419162	0.897436	0.571429
Trash	0.929385	0.96	0.944444
Glass	0.617978	0.982143	0.758621
Can	0.991803	0.75625	0.858156

2. To overcome the deficiency, we created an augmented dataset containing salt and pepper noise and varied exposure and colors, but the effect was limited.

ROS Node Optimization on Raspberry Pi We took advantage of ROS to easily integrate various nodes with different functions. We designed five nodes for image recognition, video streaming, motor control, sensor, message push in addition to the central control (Fig. 1.15). It took about 10% of the CPU computational capacity just to enable a single node on the MPU. Therefore, the CPU was almost fully loaded when the above design was used (Fig. 1.16). After simplification and reconfiguration, only three nodes for image recognition, video streaming, and central control (Fig. 1.17) were retained. The optimization allowed us to reduce the average CPU load to about 40–60% (Fig. 1.18), greatly enhancing the reliability of the system.

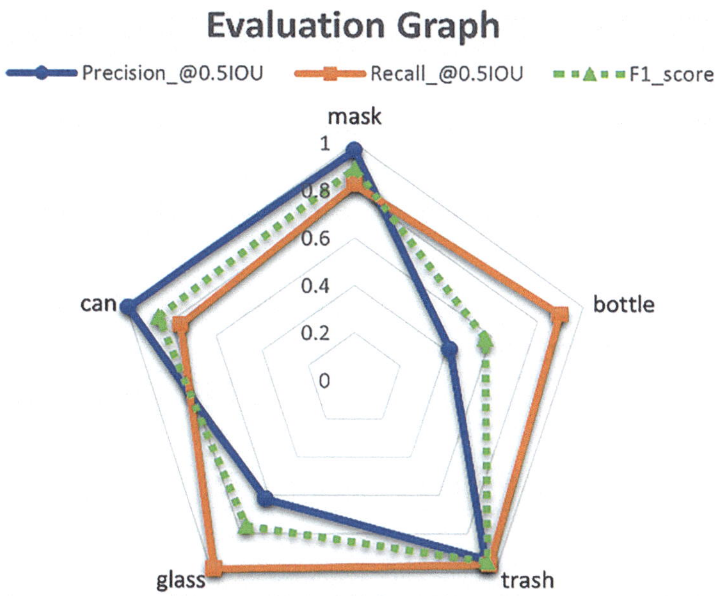

Fig. 1.14 Radar chart indicating the detection accuracy of our model

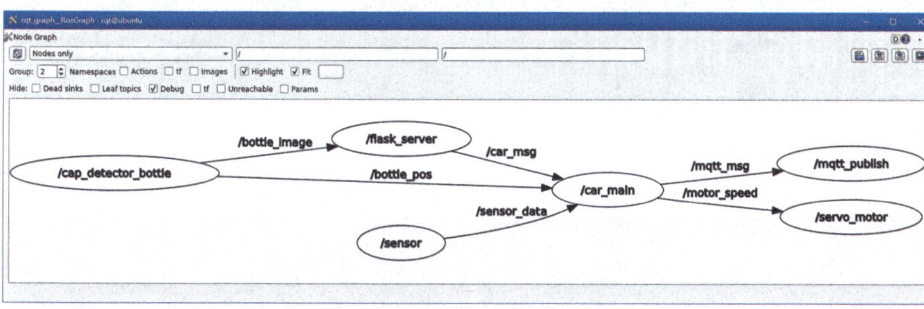

Fig. 1.15 ROS node diagram before optimization

Fig. 1.16 CPU usage before optimization

APP User Interface Optimization The APP relies on MQTT messaging and video streaming. We tested two methods of page switching. One used multiple Activities to do page switching and object re-creation. The other used multiple Fragments to

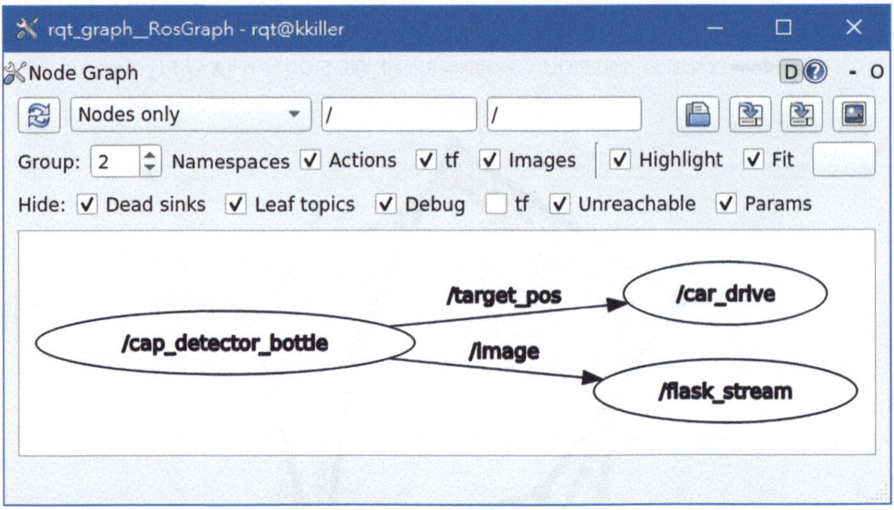

Fig. 1.17 Optimized ROS node diagram

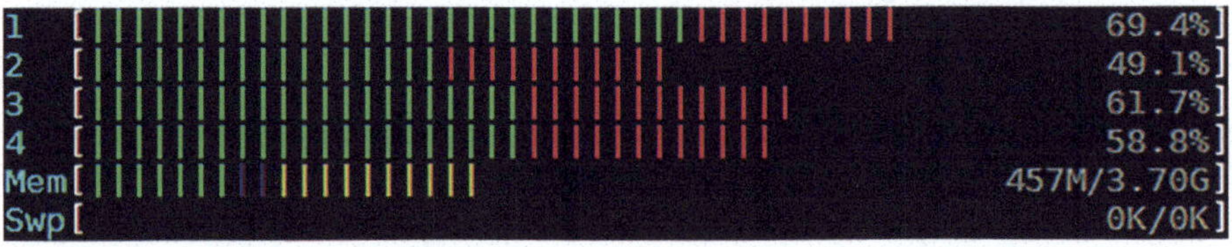

Fig. 1.18 CPU usage after optimization

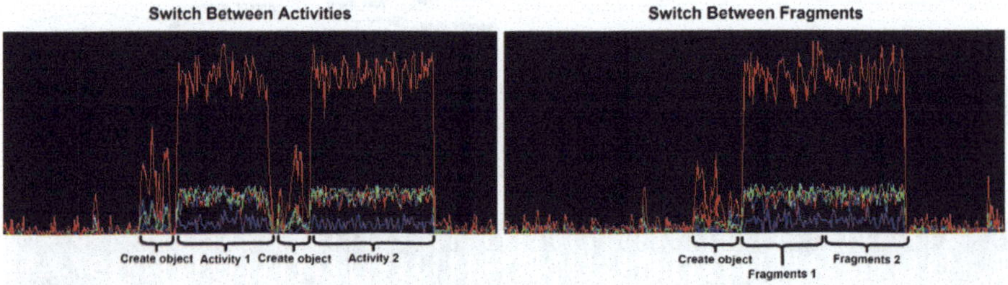

Fig. 1.19 Comparison of Activity and Fragment switching

keep the object on a single Activity through ViewModel class to do fast switching. The resource checking tool in the simulator of Android Studio shows that the way of switching pages through Fragment is almost seamless, which can realize a smoother operation experience (Fig. 1.19).

Self-Adaptive Jaw Damage-Proof Mechanism Through INA219 current measurement module, the maximum current measured is 3.2 A. Actual measurement

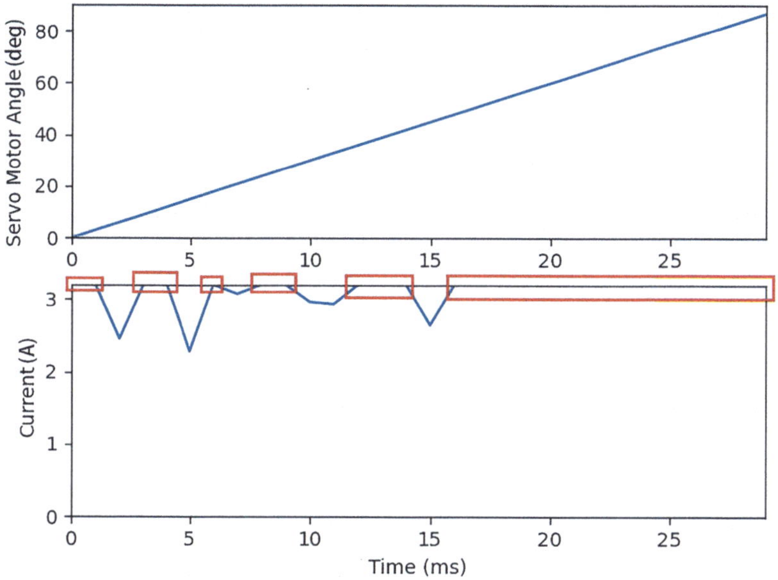

Fig. 1.20 Synchronous measurement unable to determine whether or not to clamp

showed that the average current through the servo motor of the gripping jaw was more than 2 A when rotating without a load and exceeded the maximum value when rotating with a load. The current was sampled through the I2C interface while the motor was rotating. The excessive current may cause severe damage to the motor (Fig. 1.20).

We changed the motor rotation and current measurement to a non-synchronous method. We divided the motor rotation into zones and waited for the current to stabilize every 15° (about 0.1 s) before measuring the current. Although the overall action time was longer, we could accurately determine the moment when an item was clamped (Fig. 1.21) and stop the motor in time to avoid damages to the motor.

1.6 Conclusion and Future Work

This chapter described a garbage-collecting robot that included the control board, robotic arm, gripper, motors, and wheels. The overall architecture was integrated and controlled through ROS. The client APP allows users to access various information in the database and monitor the status of the robot from a remote location. WiFi and Bluetooth wireless communication facilitate communication among the components using the lightweight MQTT protocol. Path planning and adaptive control have been incorporated to make the robot autonomous. In addition, we constructed an image detection model based on MobileNetV2. The model was

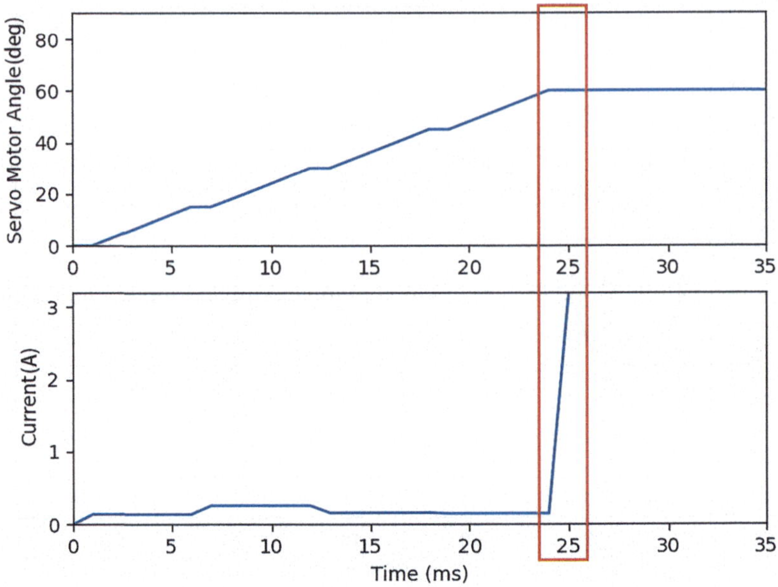

Fig. 1.21 Non-synchronous measurement to determine the proper clamping time

trained using a dataset with 3000 images containing five types of garbage. The dataset was augmented by various techniques and used to train and test the model. The average accuracy came in at 88%. Two of the garbage types exhibited lower precision, and the reason will be investigated in the future. Dependability and robustness have been taken into account in the design and implementation of the system to enhance its stability and reliability. Optimization has been applied to reduce power consumption and potential damage. Several ultrasonic rangefinders realized basic obstacle detection and avoidance. Initially, it was expected to use the sensing data to achieve simultaneous positioning and mapping (SLAM) to achieve obstacle avoidance. Alternatively, a LiDAR kit can be used. This issue will also be examined in the future.

Acknowledgments This work was supported in part by the Ministry of Science and Technology, Taiwan, under contract number MOST 110-2622-E-197-001.

References

1. Howard, A. G., Zhu, M., Chen, B., Kalenichenko, D., Wang, W., Weyand, T., Andreetto, M., & Adam, H. (2017). *MobileNets: Efficient convolutional neural networks for mobile vision applications*. Google Inc.
2. Sandler, M., Howard, A., Zhu, M., Zhmoginov, A., & Chen, L. (2018). MobileNetV2: Inverted residuals and linear bottlenecks. In *2018 IEEE/CVF conference on computer vision and pattern recognition* (pp. 4510–4520).

3. Tan, M., & Le, Q. V. (2019). EfficientNet: Rethinking model scaling for convolutional neural networks. *International Conference on Machine Learning*, 6105–6114.
4. Dai, J., Li, Y., He, K., & Sun, J. (2016). RFCN: Object detection via region-based fully convolutional networks. In *Advances in neural information processing systems* (pp. 379–387).
5. Chen, J., Han, G., Guo, S., & Diao, W. (2018). FragDroid: Automated user interface interaction with activity and fragment analysis in android applications. In *48th annual IEEE/IFIP international conference on dependable systems and networks (DSN)* (pp. 398–409). https://doi.org/10.1109/DSN.2018.00049
6. Xu, Y., & Li, H. (2010). Research on interconnection and correspondence between relational database and embedded database SQLite. *International Conference on Communications and Mobile Computing, 1*, 279–283. https://doi.org/10.1109/CMC.2010.188
7. Kong, X., Fan, B., Nie, W., & Ding, Y. (2016). Design on mobile health service system based on android platform. In *IEEE advanced information management, communicates, electronic and automation control conference (IMCEC)* (pp. 1683–1687). https://doi.org/10.1109/IMCEC.2016.7867504
8. Megalingam, R. K., Katta, N., Geesala, R., Yadav, P. K., & Rangaiah, R. C. (2018). Keyboard-based control and simulation of 6-DOF robotic arm using ROS. In *4th international conference on computing communication and automation (ICCCA)* (pp. 1–5). https://doi.org/10.1109/ccaa.2018.8777568
9. Deng, H., Xiong, J., & Xia, Z. (2017). Mobile manipulation task simulation using ROS with MoveIt. In *IEEE international conference on real-time computing and robotics (RCAR)* (pp. 612–616). https://doi.org/10.1109/RCAR.2017.8311930
10. Braier, Z., & Klouček, P. (2015). System of measurement and evaluation of AC servo motor's mechanic, electric and control quantities. In *IEEE international workshop of electronics, control, measurement, signals and their application to mechatronics (ECMSM)* (pp. 1–5). https://doi.org/10.1109/ECMSM.2015.7208691
11. Ting, C.-Y. (2016). *The investigation of the Robot's dynamic path planning*. Master's thesis, Department of Information Management, Xuanzang University, Hsinchu, Taiwan.
12. Lai, J. (2017). *Proximity sensing using multiple 3D depth sensors in a robotic operating system*. Master's thesis, Department of Electrical Engineering, National Sun Yat-sen University, Kaohsiung, Taiwan.
13. Tang, C.-Y. (2018). *Design and verification of automatic return charging mechanism for floor sweeping robots*. Master's thesis, Department of Information Engineering, National Yunlin University of Science and Technology, Yunlin, Taiwan.
14. Yoo, H., Wee, S., & Son, Y. (2019). Buck converter for noise-isolating power supplies. In *IEEE 2nd international conference on electronics and communication engineering (ICECE)* (pp. 403–406). https://doi.org/10.1109/ICECE48499.2019.9058501
15. Kim, Y. H., Nam, S. H., & Park, K. R. (2021). Enhanced cycle generative adversarial network for generating face images of untrained races and ages for age estimation. *IEEE Access*, 6087–6112. https://doi.org/10.1109/ACCESS.2020.3048369

Chapter 2
Cybersecurity Data Science: Concepts, Algorithms, and Applications

Wei Lu

2.1 Introduction

Cybersecurity is an ability to protect or defend the use of cyberspace from cyberattacks. The four most common goals provided by cybersecurity technologies for any users/entities in cyberspace are confidentiality, authentication, message integrity, and access and availability [1]. Confidentiality ensures that the information is inaccessible to unauthorized people, and it is commonly enforced through encryption, IDs and passwords, two-factor authentication, and additional defensive strategies in which only sender and the intended receiver should understand message contents. In the context of confidentiality, the sender encrypts the message and the receiver decrypts the message. A typical example of confidentiality is student grade information that is an asset whose confidentiality is important to students, and only be available to students, their parents, and employees that require the information to do their job. Message integrity is to safeguard information and systems from being modified by unauthorized people, thereby ensuring that the protected data is accurate and trustworthy. Considering, for example, a hospital patient's allergy information stored in a database and trusted by the doctor, the integrity is broken when an employee (e.g., a nurse) who is authorized to view and update this information deliberately falsified the data to cause harm to the hospital. Then, the database needs to be restored to a trusted basis quickly, and trace the error back to the person responsible. In the context of message integrity, sender and receiver want to ensure messages are not altered in transit, or afterwards

W. Lu (✉)
Department of Computer Science, Keene State College, The University System of New Hampshire, Keene, NH, USA
e-mail: wlu@usnh.edu

I. Woungang, S. K. Dhurandher, *4th International Conference on Wireless, Intelligent and Distributed Environment for Communication*, Lecture Notes on Data Engineering and Communications Technologies 94, https://doi.org/10.1007/978-3-030-89776-5_2

without detection. Availability is to ensure that authorized people have access to the information when needed; this includes rigorously maintaining all systems, keeping them current with upgrades, using backups to safeguard against disruptions or data loss, to name a few. For example, a system providing authentication services for critical systems, applications, and devices. An interruption of service results in the inability for customers to access computing resources and staff to access the resources they need to perform critical tasks. The loss of the service translates into a large financial loss in lost employee productivity and potential customer loss. In the context of availability, services must be accessible and available to users 24 h 7 days.

The early cybersecurity mechanisms dedicated exclusively to prevention are highly insufficient mainly because current Internet activities have shown a dramatic increase in the number of computer and network attacks [2]. It was estimated that cybercrime is the second most-strongly connected global risk, and the cost of its damage would be equivalent to the GDP of the world's third largest economy in 2021 [3]. Under current circumstances, password-based authentication and access control mechanisms, which represent the cornerstone of traditional protection systems, can easily be circumvented using widely available exploits [4]. As a result, the concept of cybersecurity data science is proposed to supplement traditional prevention-based security mechanisms.

In the rest of this short chapter, we introduce the basic concepts of cybersecurity data science in Sect. 2.2 including what is cybersecurity, information security and network security, and what are their main differences, what is data science, and what is cybersecurity data science. In Sect. 2.3, we review an unsupervised machine learning algorithm [5], and then in Sect. 2.4, we present how to use the proposed unsupervised machine learning algorithm for detecting zero-day attacks, a typical case study on cybersecurity data science. Section 2.5 makes some concluding remarks and discusses future work.

2.2 Concepts of Cybersecurity Data Science

Cybersecurity, information security and network security are three terms that are often used interchangeably, even among some of those in the security field. However, they are not exactly the same. Each of them addresses different kinds of security. Understanding each term, what it means, and the difference among the three, i.e., what are they, how are they different, and why are these terms so often confused, is basic and essential for any organization that is investing in a proper security framework.

We have known that cybersecurity is an ability to protect or defend the use of cyberspace from cyberattacks, while computer security is a set of measures and controls that ensure confidentiality, integrity, and availability of the information processed and stored by a computer [6]. This term has been replaced by the term cybersecurity. On the other hand, information security is the protection of

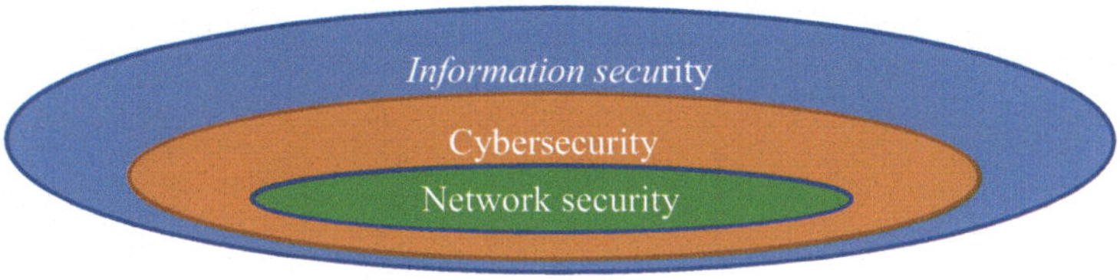

Fig. 2.1 Information security vs. cybersecurity vs. network security

information and information systems from unauthorized access, use, disclosure, disruption, modification, or destruction in order to provide confidentiality, integrity, and availability, and network security represents the process of taking physical and software preventative measures to protect the underlying networking infrastructure from unauthorized access, misuse, malfunction, modification, destruction, or improper disclosure, thereby creating a secure platform for computers, users, and programs to perform their permitted critical functions within a secure environment.

As illustrated in Fig. 2.1, information security is, broadly, the practice of securing your data, no matter its form, and cybersecurity is a subset of information security that deals with protecting an organization's internet-connected systems from potential cyberattacks; moreover, network security is a subset of cybersecurity that is focused on protecting an organization's IT infrastructure from online threats. Although the terms are often used in conjunction with one another, cybersecurity is considered to be the broader discipline as long as data originates in a digital form.

2.2.1 What Is Data Science?

Generally speaking, data science is an interdisciplinary field that uses scientific methods, processes, algorithms, and systems to extract knowledge and insights from many structural and unstructured data. As illustrated in Fig. 2.2, the Data Science Venn Diagram includes six components, namely Data Hacking Skills, Math and Statistics Knowledge, Machine Learning, Traditional Research, Danger Zone, and Substantive Expertise (domain knowledge) [7].

- *Data Hacking Skills*: Data hacker/engineer with skills to acquire and clean data, thinking algorithmically, being able to manipulate data using command line, understanding vectorized operations, to name a few.
- *Math and Statistics Knowledge*: Extract insight from the data, at least needs to know what is least squares regression to be a competent data scientist.
- *Substantive Expertise (Domain Knowledge)*: Substantive expertise is essential and is built upon domain knowledge, because data science is about discovery and building knowledge, which at least requires some motivating questions about the world and hypotheses that can be brought to data and tested with statistical methods.

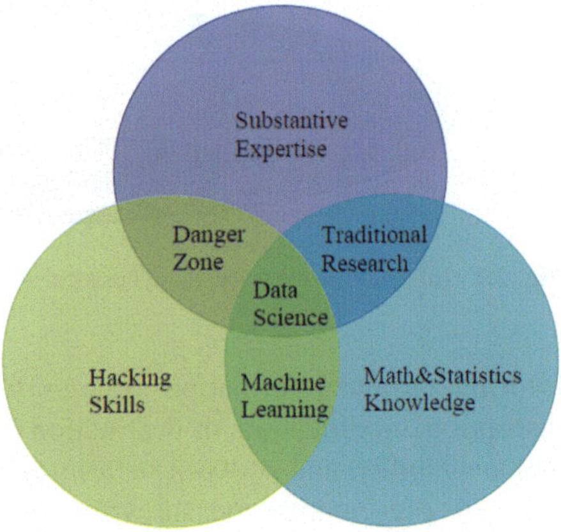

Fig. 2.2 The data science Venn diagram [7]

- *Machine Learning*: Data plus and math/statistics get us to machine learning.
- *Traditional Research*: Domain knowledge and math/statistics get you traditional research.
- *Danger Zone*: Data plus domain knowledge is sparsely populated and is the most problematic of the diagram, could produce a lot of false interpretation without learning some math and statistics along the way.

As a result, data science is an interdisciplinary field that combines Data Hacking Skills, Math and Statistics Knowledge, and Substantive Expertise (Domain Knowledge).

2.2.2 What Is Cybersecurity Data Science?

Cybersecurity data science is an interdisciplinary field that combines data hacking skills, math and statistics knowledge, and substantive expertise in cybersecurity including in particular one or more of the following subjects:

- SIEM development
- Insider threat detection
- Computer and network forensics
- Security metrics; governance, risk, and compliance
- Risk modeling; fraud and loss analytics
- Advanced threat mitigation; malware analysis

2.3 Algorithms for Cybersecurity Data Science

The EM algorithm is widely used to estimate the parameters of Gaussian Mixture Model (GMM) [8]. GMM is based on an assumption that the data to be clustered are drawn from one of several Gaussian distributions. It is suggested that Gaussian mixture distributions can approximate any distribution up to an arbitrary accuracy, as long as a sufficient number of components are used. Consequently, the entire data collection is seen as a mixture of several Gaussian distributions, and their corresponding probability density functions can be expressed as a weighted finite sum of Gaussian components with different parameters and mixing proportions.

The conditional probability in EM describes the likelihood that data points approximate a specified Gaussian component. The greater the value of conditional probability for a data point belonging to a specified Gaussian component, the more accurate the approximation is. As a result, data points are assigned to the corresponding Gaussian components according to their conditional probabilities. However, in some cases, there exist some data points whose conditional probability of belonging to any component of a GMM is very low or close to zero. These data are naturally seen as the outliers or noisy data. All the outlier data will be deleted or considered as unknown (aka zero-day) attacks during anomaly detection, and their attacking probability is set to 1.0. Table 2.1 illustrates a detailed EM-based clustering algorithm in which C_m stands for the clustering results.

In order to apply the EM-based clustering technique for detecting zero-day attacks, we make two basic assumptions: (1) the input data points are composed of two clusters, namely anomalous cluster and normal cluster; (2) the size of the anomalous cluster is always smaller than the size of the normal cluster. Consequently, we can easily label the anomalous cluster according to the size of each cluster. The attack probability for each data point is equal to the conditional probability of corresponding data point belonging to the anomalous cluster, which is defined as follows:

$$p = p_{r-1}\left(C_{\text{anomalous}} | x_n\right)$$

where x_n is the data point; $C_{\text{anomalous}}$ is the anomalous cluster; $p_{r-1}(C_{\text{anomalous}} | x_n)$ is the conditional probability of x_n belonging to anomalous cluster $C_{\text{anomalous}}$.

Table 2.1 EM-based clustering algorithm

Function EMCA (data) **returns**
clusters C_m and posterior probability $p_r(i | x_n)$
$C_m = \phi$, $1 \pounds m \pounds k$, k is the number of clusters
Call EM (data);
For $1 \pounds m \pounds k$, $1 \pounds n \pounds N$
If ($p_{r-1}(m | x_n) = \max(p_{r-1}(m | x_n))$)
Then assign x_n to $\mathrm{C_m}$
Return C_m, $m = 1, 2..., k$

2.4 Applications for Cybersecurity Data Science

In order to apply the proposed EM-based clustering algorithm for detecting zero-day attacks in cybersecurity data science as a case study, we develop an effective and efficient clustering framework for online unsupervised anomaly detection of unknown intrusions. The proposed detection framework consists of a feature extraction technique based on the number of network flows and packets and an EM-based unsupervised machine learning algorithm. Figure 2.3 depicts the general architecture of our framework, which consists of three main stages as follows:

1. Feature analysis: During this phase, a number of flows and packets are generated from standard packet information, allowing extraction of salient and useful domain knowledge, as well as significant reduction of the dimensionality in the feature space.
2. Clustering: The features computed during the previous step are clustered using the EM-based clustering algorithm.
3. Cluster analysis and intrusion decision: Analyzing the clustering outcomes using heuristics leads to a final classification of corresponding data as either unknown novel attacks or nonintrusive data.

The major goal of feature analysis is to select and extract robust network features that have the potential to discriminate anomalous behaviors from normal network activities. Since most current network intrusion detection systems use network flow data (e.g., netflow, sflow, ipfix) as their information sources, we focus on features in terms of flows.

FlowCount A flow consists of a group of packets going from a specific source to a specific destination over a time period. There are various flow definitions so far, such as netflow, sflow, ipfix, to name a few. Basically, one network flow should at least include source IP/port, destination IP/port, protocol, number of bytes, and number of packets. Flows are often considered as sessions between users and services. Since attacking behaviors are usually different from normal user activities, they may be detected by observing flow characteristics.

PacketCount The average number of packets in a flow over a time interval. Most attacks happen with an increased packet count. For example, Distributed Denial of

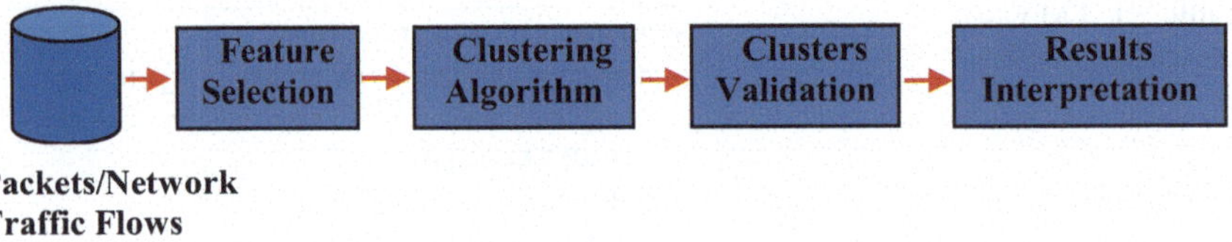

Fig. 2.3 A typical cybersecurity data science application framework

Table 2.2 List of features

Features	Description
f_1	Number of TCP flows per minute
f_2	Number of UDP flows per minute
f_3	Number of ICMP flows per minute
f_4	Average number of TCP packets per flow over 1 m
f_5	Average number of UDP packets per flow over 1 m
f_6	Average number of ICMP packets per flow over 1 m
f_7	Average number of bytes per TCP flow over 1 m
f_8	Average number of bytes per UDP flow over 1 m
f_9	Average number of bytes per ICMP flow over 1 m

Service (DDoS) attacks often generate a large number of packets in a short time in order to consume the available resources quickly.

ByteCount The average number of bytes in a flow over a time interval. Through this metric, we can identify whether the network traffic consists of large size packets or not. Some previous Denial of Service (DoS) attacks use maximum packet size to consume the computation resources or to congest data paths, such as the well-known ping of death (pod) attack.

Based on the above three metrics, we define a set of features to describe entire traffic behavior on networks. Let F denote the feature space of network flows, a nine-dimensional feature vector $f\in$ can be represented as $\{f_1, f_2, \ldots, f_9\}$, where the meaning of each feature is explained in Table 2.2.

We evaluate the application framework with the full 1999 DARPA intrusion detection dataset. In particular, we conduct a comprehensive analysis for network traffic provided by the dataset and identify the intrusions based on each specific day. Since most current existing network intrusion detection systems use network flow data (e.g., netflow, sflow, ipfix) as their information sources, we covert all the raw TCPDUMP packet data into flow-based traffic data by using the public network traffic analysis tools (e.g., editcap [9], tshark [10]), similarly to the 1999 KDDCUP dataset [11] in which the 1998 DAPRA intrusion detection dataset [12] has been converted into connection-based dataset. Although the 1998 and 1999 DARPA dataset were criticized in [13, 14] due to the methodology for simulating the actual network environment, they are the widely used and acceptable benchmark for the current intrusion detection research.

During the evaluation, the results are summarized and analyzed in three different categories, namely how many attack instances are detected by each feature, how many attack types are detected by each feature, and how many attack instances are detected for each attack type. We do not use the traditional receiver operating characteristic (ROC) curve to evaluate our approach and analyze the tradeoff between the false-positive rates and detection rates because ROC curves are often misleading and incomplete. Compared to most, if not all, other evaluations with the 1999 DARPA dataset, our evaluation covers all types of attacks and all days'

Table 2.3 Performance of all nine features over 9 days evaluation for EM clustering

Features	Average DR (%)	Average FPR (%)	Ratio of Avg. DR to Avg. FPR
F1	39.83	81.84	0.487
F2	52.22	84.04	0.621
F3	32.25	84.14	0.383
F4	12.0	89.03	0.135
F5	51.8	85.74	0.604
F6	32.25	84.17	0.383
F7	3.2	82.92	0.0386
F8	49.26	84.19	0.585
F9	32.25	84.14	0.383

network traffic, and thus, we consider our evaluation as a comprehensive analysis for network traffic in the 1999 DARPA dataset.

The scoring coefficient is set up according to the detection rate (DR) and the false-positive rate (FPR) for the developed system over a long history. The higher the DR, the better the performance of the system; the lower the FPR, the better the system performance. Therefore, the ratio of DR to FPR is used to measure the performance of our proposed framework. We evaluate our system with the nine features and 9 days DARPA testing data on weeks 4 and 5. The evaluation results are summarized and analyzed in three different categories described in above. Table 2.3 illustrates the average value of DR, FPR, and the ratio of DR to FPR for each feature over those 9 days. From Table 2.3, we can see that the top three features that are extremely useful for detecting zero-day attacks are number of UDP flows per minute, average number of UDP packets per flow over 1 min, and average number of bytes per UDP flow over 1 min.

2.5 Conclusions and Future Work

In this short chapter, we discuss the essential concepts of cybersecurity data science and its typical algorithms and applications in detecting novel network attacks. The concept of cybersecurity data science is not new and can be originally dated back to the 1980s in the seminal report of Denning [15]. Denning assumed that security violations could be detected by inspecting abnormal system usage patterns from the audit data. Deviations from normal behavior patterns are flagged systematically as intrusions. The implementations of early anomaly detection techniques were based on self-learning. Knowledge about normal behaviors of subjects was automatically formed through training. Thus, according to whether the learning process is supervised or unsupervised, the anomaly detection schemes are naturally classified into two categories: unsupervised and supervised.

Supervised anomaly detection schemes depend on labeled training datasets, making the intrusion detection process error-prone, costly and time consuming.

Any mistake in labeling the training data may lead to decreased performance of the detector. Unsupervised anomaly detection schemes allow training based on unlabeled datasets, facilitating online learning and improving detection accuracy. By facilitating online learning, unsupervised approaches provide higher potential to find novel attacks, which are not always included in the training data [16–18]. By removing the need to label the dataset, unsupervised approaches carry greater potential for detection accuracy, and they, however, also carry greater computing overheads. In the future, we will have supervised learning, in particular, a traditional perceptron learning algorithm, and unsupervised learning approaches integrated into the proposed cybersecurity data science application framework in order to bridge the knowledge gap between fields so that legal practitioners, law enforcement agencies, policy makers, and economists can make wise choices when dealing with societal problems related to cybersecurity as well as its broad economic impact.

Acknowledgments This research was supported in part by funding from a Keene State College Faculty Development Grant.

References

1. Stallings, W. (2019). *Cryptography and network security: Principles and practice* (8th ed.). Pearson. ISBN-13: 978-0135764183.
2. Ghorbani, A. A., Lu, W., & Tavallaee, M. (2010). Network attacks. In *Network intrusion detection and prevention. Advances in information security* (Vol. 47). Springer.
3. The Global Risks Report 2020. (2021). https://www.weforum.org/reports/the-global-risks-report-2020
4. Lu, W., & Traore, I. (2005). An unsupervised approach for detecting DDOS attacks based on traffic-based metrics. In *2005 IEEE pacific rim conference on communications, computers and signal processing* (pp. 462–465).
5. Lu, W., & Traore, I. (2005). Determining the optimal number of clusters using a new evolutionary algorithm. In *17th IEEE international conference on tools with artificial intelligence (ICTAI'05).*
6. National Institute of Standards and Technology. (2015). https://csrc.nist.gov/glossary/
7. The Data Science Venn Diagram. (2013). http://drewconway.com/zia/2013/3/26/the-data-science-venn-diagram
8. Lu, W., & Traore, I. (2005). A new evolutionary algorithm for determining the optimal number of clusters. In *International conference on computational intelligence for modelling, control and automation and international conference on intelligent agents, web technologies and internet commerce* (Vol. 2005, pp. 648–653).
9. Editcap. https://www.wireshark.org/docs/wsug_html_chunked/AppToolseditcap.html
10. Tshark. https://www.wireshark.org/docs/man-pages/tshark.html
11. KDDCUP. http://archive.ics.uci.edu/ml/datasets/KDD+Cup+1999+Data
12. *The 1998 DARPA intrusion detection dataset.* https://www.ll.mit.edu/r-d/datasets/1998-darpa-intrusion-detection-evaluation-dataset#:~:text=Intrusion%20detection%20systems%20were%20delivered,real%20time%20during%20normal%20activities
13. Mahoney, M. V., & Chan, P. K. (2003). An analysis of the 1999 DARPA/Lincoln Laboratory evaluation data for network anomaly detection. In *Proceedings of the 6th international symposium on recent advances in intrusion detection* (pp. 220–237).

14. McHugh, J. (2000). Testing intrusion detection systems: A critique of the 1998 and 1999 DARPA intrusion detection system evaluations as performed by Lincoln Laboratory. *ACM Transactions on Information and System Security, 3*(4), 262–294.
15. Denning, D. E. (1987). An intrusion detection model. *IEEE Transactions on Software Engineering, SE-13*(2), 222–232.
16. Lu, W., & Xue, L. (2014). A heuristic-based co-clustering algorithm for the internet traffic classification. In *2104 28th international conference on advanced information networking and applications workshops* (pp. 49–54). https://doi.org/10.1109/WAINA.2014.16
17. Garant, D., & Lu, W. (2013). Mining botnet behaviors on the large-scale web application community. In *2013 27th international conference on advanced information networking and applications workshops* (pp. 185–190). https://doi.org/10.1109/WAINA.2013.235
18. Lu, W., & Traore, I. (2008). Unsupervised anomaly detection using an evolutionary extension of k-means algorithm. *International Journal on Information and Computer Security, 2*(2), 107–139.

Chapter 3
Comparison of Task Scheduling Algorithms for Traffic Surveillance Application Using Fog Computing

Mluleki Sinqadu, Zelalem Sintayehu Shibeshi, and Khuram Khalid

3.1 Introduction

Recent advances in technologies, particularly digital technologies, have led to the presentation of another term known as the Fourth Industrial Revolution (4IR), which is promoted as the answer to everything from accumulations in education and improvement skills to organizations reducing expenses and serving their clients better. The Internet of Things (IoT) is one of the pillars of 4IR, characterized as the interaction among automation and data interchange. IoT is the prominent technology that creates the possibilities of communication between physical objects and the Internet [1]. Because of its advantages, many IoT-enabled systems and applications have been created. The applications are particularly important in domains such as smart homes, medical services, smart transportation, to give some examples. At the centre of the assessment and approval of these systems is the capacity to plan proper application models and the analysis of fog computing, which consists of fog devices, IoT devices and cloud servers. Taking into account that implementation of the fog environment can be very costly, researchers have decided to develop simulators that can be used to test real-life scenarios of IoT applications and cloud-based applications in an adaptable way.

M. Sinqadu (✉) · Z. S. Shibeshi
Department of Computer Science, University of Fort Hare, Alice, South Africa
e-mail: zshibeshi@ufh.ac.za

K. Khalid
Department of Computer Science, Ryerson University, Toronto, ON, Canada
e-mail: khuram.khalid@ryerson.ca

I. Woungang, S. K. Dhurandher, *4th International Conference on Wireless, Intelligent and Distributed Environment for Communication*, Lecture Notes on Data Engineering and Communications Technologies 94, https://doi.org/10.1007/978-3-030-89776-5_3

iFogSim [1] is an illustration of such a system, which has been used for designing models for fog/edge computing environment and running experiments. iFogSim has been developed by inheriting some basic CloudSim classes [2]. The design of Fog computing in iFogSim, alongside with intelligent surveillance model using a network of distributed cameras, is depicted in [1]. In [3], an API for fog applications was presented, and a Closed-Circuit TV (CCTV)-based vehicle tracking framework was demonstrated using the proposed API, taking note of that CCTV is a self-contained framework made out of cameras, recorders for surveillance. This chapter investigates how iFogSim is used to simulate and evaluate the performance of the proposed traffic surveillance application like the one presented in [4]. We did this by following the fog-based application model presented in [5] for a case scenario of a traffic surveillance application. In the work introduced in this framework, the model is improved with new components, adding new functionalities in the design of the following vehicle tracking process. However, in this chapter, we evaluate the same model using task scheduling algorithms on metrics that have been introduced in Sect. 3.4.

This chapter is organized as follows. In Sect. 3.2, some related works are presented. In Sect. 3.3, a brief introduction of scheduling algorithms that are used for evaluation is provided. In Sect. 3.3, we discuss the evaluation metrics that are used to test our model. In Sect. 3.5, the design of the proposed fog-based application model is described, along with justification for using iFogSim. In Sect. 3.6, the performance analysis of the Traffic Surveillance application model is assessed by simulations, using scheduling algorithms. Finally, Section 3.6 is the conclusion of this chapter.

3.2 Related Work

In real-time fog applications, before a task is executed, the fundamental question one should ask would be whether it should be cloud or fog [6]. For real-time applications that require emergence response, fog is always chosen over the cloud for processing as it can provide cloud resources closer to devices. The following related works review how fog computing can benefit IoT application and also reducing blocking while minimizing latency. In this section, we present works related to our project which demonstrate how different applications are evaluated using task scheduling algorithms.

Intharawijitr et al. [7] presented different policies that can be used to map tasks for fog devices while minimizing latency. The first policy selects which fog device to host the current task, whereas the second policy selects the fog device to reduce overall latency, and the last policy decides which fog device to increase resource usage. Results show that to avoid task blocking, the algorithm that prioritizes latency must be used for mapping tasks.

In [8], the authors proposed a strategy for task scheduling that checks the organization of clusters of fog devices and how the load can be balanced among the cellular network. The strategy allocates resources at the serving cell according to an ordered list of tasks and arranges clusters to meet the requirements of the tasks not yet served. Results indicate that the strategy works better in a cellular network when compared to static clustering.

A model that ensures minimum resource usage for fog resource usage of various IoT devices is presented in [9]. This model uses an approximation of resource usage. Moreover, it reflects the history of resource usage and starts with the resource allocation to frequent users.

On the other hand, Aazam and Huh [9] explored the issue of reasonable task offloading among fog devices in a 5G fog network. They demonstrated a task scheduler that takes into account task latency and energy usage. The fairness of using a fog system to reduce task latency is also demonstrated. The findings show that task scheduling among fog devices can reduce task execution time.

In [10], Deng et al. discussed the difference between power consumption and delay in a cloud-fog system. The authors proposed a model that solves the problem with resource allocation for power consumption and delays on a cloud-fog distribution network. To reduce the energy consumed by a fog layer must increase convex usage for a certain input workload of tasks, while a heuristic is used to improve the energy usage of a cloud. Moreover, the authors try to maximize the energy usage of communications. Results show that by giving away few resources, bandwidth can be saved and hence latency is minimized.

Pham and Huh [11] proposed a task scheduler to advance the efficiency of the method of offloading applications for a large scale. The objective is to achieve a decent compromise among span and financial expense. The scheduler sorts the tasks according to their span and duration of execution; then chooses the appropriate fog device to execute each task in the list.

In [12], Zeng et al. proposed a formulation to choose whether the processing of tasks ought to be completed on client devices or edge devices. The objective is to reduce the time taken to finish the tasks considering task scheduling. The time taken to finish each task incorporates computational time and data transmission time. Results demonstrate that the presented algorithms show incredible performance on processing data to edge devices for different task appearance rates and customer handling rates.

All the studies discussed above are related to our work, which is the implementation of the fog-based Traffic Surveillance Application model. In the following section, we will discuss the selected algorithms to evaluate our model based on four metrics and the design of our proposed model.

3.3 Selected Algorithms

The scheduling algorithms used to evaluate our proposed model are selected based on work done in [13]. Those algorithms are selected because they have been used successfully in cloud computing for evaluating different applications [14, 15]. In this work, those algorithms are applied to a distributed network of smart cameras. The first algorithm, First Come First Serve (FCFS), is a simple scheduling algorithm that executes the queued tasks and processes in the order in which they are received. It is the most straightforward CPU scheduling algorithm. The second algorithm is Short Job First (SJF), which selects the task with the shortest execution time as the next task to run. Preemptive or non-preemptive scheduling algorithms are available. It greatly decreases the total time spent waiting for other processes to complete. The third algorithm used is Round Robin (RR), in which each ready task runs in a cyclic queue for a predefined amount of time. This algorithm also allows for process execution without starvation. Finally, we consider the Generalized Priority (GP) scheduling algorithm, a popular option for real-time applications which ensures that the processor executes the task with the highest priority of all tasks that are currently ready to execute at any given time.

3.3.1 Motivation for Selected Algorithms

The scheduling algorithms discussed above have been evaluated in cloud computing before and proven to provide efficiency. Likewise, fog computing applications also require low latency while carrying out tasks efficiently. These algorithms are now used to evaluate a vehicular network application model that requires low latency and efficiency. To find a well-performing algorithm, different metrics must be used to evaluate all the algorithms and a comparison can be done. The selected algorithms have shown good scheduling of tasks on cloud computing applications; therefore, they are more suitable to evaluate the Traffic Surveillance Application. Our comparison of the studied algorithms reveals how efficient and fast each algorithm satisfies the application.

FCFS algorithm can be used in this model as it schedules the tasks based on arrival queue, which in this case are vehicle received detected from raw video streams which are discussed in Sect. 3.5. SJF is another algorithm that we consider suitable for vehicular networks as it processes the tasks with the shortest execution time as the next task to run. This algorithm is also fitting in this study as it helps in reducing accidents and process vehicles which are faster and likely to break the road rules. Another algorithm we are considering is RR, this algorithm runs each ready task in a cyclic queue; it is also suitable for a vehicular network to ensure that it gets the predefined time and also allow processing of the task without starvation. The last algorithm is GP, which is popular for real-time applications to ensure that tasks with high priority are executed at any given time. This algorithm is also suitable for a vehicular network to ensure the processing of tasks in real-time. All these algorithms are chosen with suitability with the vehicular network, but there is only one algorithm that can be chosen as most suitable, and it is selected based on the demonstration of good performance in all given metrics discussed in Sect. 3.6. Therefore, it will be the most suitable algorithm for a real-time IoT application.

In the following section, we will report on the metrics that will be used to evaluate our model considering the aforementioned scheduling algorithms.

3.4 Evaluation Metrics

In a fog distributed network, the performance evaluation is done based on latency, energy consumption, execution time and network usage. These evaluation metrics are briefly explained below:

Average Latency
Average Latency refers to the delay that happens when data is transmitted between surveillance cameras and the Cloud which can be calculated as shown in Eq. (3.1).

$$\text{Average Latency} = \text{CC–ET} \tag{3.1}$$

where CC is the CloudSim Clock and ET is the Emitting Time of a tuple. ET is calculated by the transmission time of a module to another module.

Energy Consumption
The energy consumption for the full topology of the network is calculated using Eq. (3.2).

$$\text{Energy} = \text{CEC} + (\text{NT} - \text{LUUT}) * \text{HP}. \tag{3.2}$$

The "energy consumption is calculated by taking the power of all components in a certain time frame of execution, where CEC is the current energy consumption, NT is the network time, LUUT is the last utilization update time, and HP is the host power in LU."

Execution Time
Execution Time refers to the time taken by the tuple on consuming fog device resources. It can be mathematically calculated using Eq. (3.3):

$$\text{Execution Time} = \text{CT–SST} \tag{3.3}$$

where CT represents the current time and SST denotes the simulation start time.

Network Usage
"Network Usage refers to the amount of network bandwidth consumed while running simulation. It can be mathematically described using" Eq. (3.4):

$$\text{Network Usage} = (\text{TL} * \text{TC})\,/\text{MST} \tag{3.4}$$

where TL is the total latency and TS is the total size of the tuple, respectively; and MST is the maximum simulation time.

3.5 Design of a Proposed Traffic Surveillance Application Model for Fog Computing

In this section, we discuss the use case and design of our proposed application model.

3.5.1 Fog Computing Application Model Topology

In Fig. 3.1, a fog topology of the Traffic Surveillance Application model is presented which shows all the devices used.

Figure 3.1 illustrates how each process depends on one another in the fog topology tree from cloud to fog devices. The router in the topology represents the fog device while mobile-0 and mobile-1 represent the smart cameras physical infrastructure. Lastly, we have motor-0 and motor-1 referring to smart cameras and the pan–tilt–zoom (PTZ) actuation to vehicle detection.

3.5.2 Traffic Surveillance Application Model Design

Figure 3.2 shows our proposed fog-based application model with all five modules which have different tasks. These modules are interconnected by tuples which help in transmitting the data from one module to another.

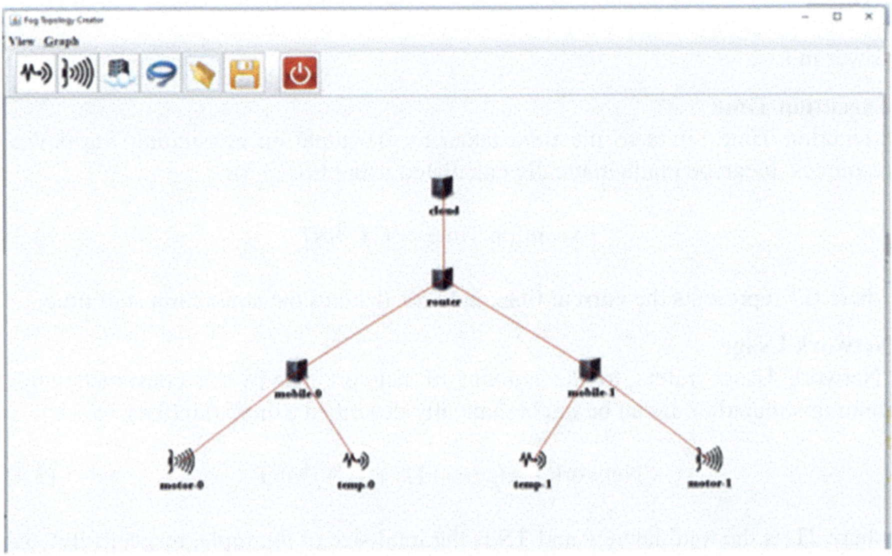

Fig. 3.1 Fog computing application model topology

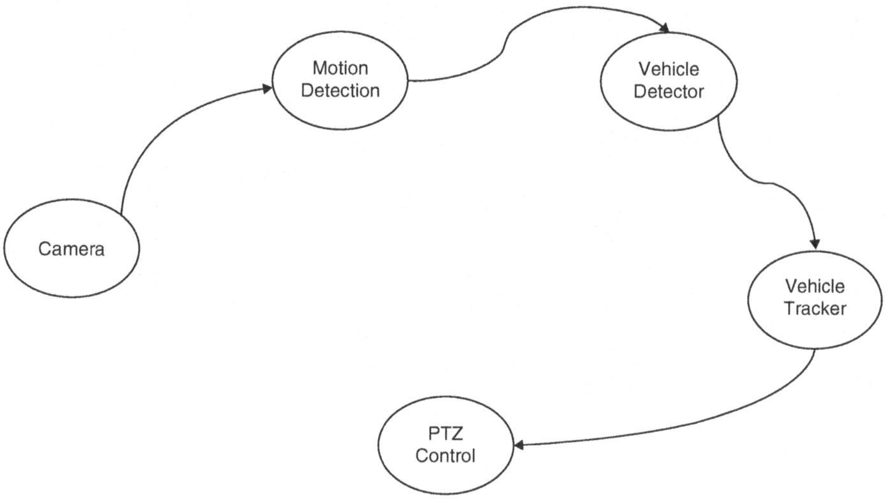

Fig. 3.2 Proposed fog-based application model

As mentioned earlier, the Traffic Surveillance application model relies on surveillance cameras that can detect moving vehicles. This application requires sufficient processing power for a good experience. Our surveillance application model consists of five modules as shown in Fig. 3.2. "These modules are motion detector, vehicle detector, vehicle tracker, user interface and pan–tilt–zoom (PTZ) controller. The motion detector module retrieves the raw video streams captured by a smart camera and sends the video to the vehicle detector module. The vehicle detector module detects the moving vehicles and sends vehicle identification and the current position of the vehicle to the vehicle tracker. The vehicle tracker obtains the coordinates of the tracked vehicles, computes an ideal PTZ configuration of all the cameras surveilling the area and then forwards the command to the PTZ control module. The PTZ module, placed in each smart camera, rotate the smart camera based on the PTZ parameters received from the vehicle tracker module. Lastly, the user interface module forwards a portion of the video streams containing each tracked vehicle to the user's device. Since it is assumed that each smart camera is equipped with a PTZ control module, our application model of the Traffic Surveillance application is composed of four modules."

3.5.3 Sequence Diagram of the Proposed Application Model

Figure 3.3 demonstrates a sequence diagram of our proposed model and how objects interact in the arranged time sequence.

The sequence diagram in Fig. 3.3 depicts the interaction of the main objects of the Traffic Surveillance Model and how operations are carried out. This interaction

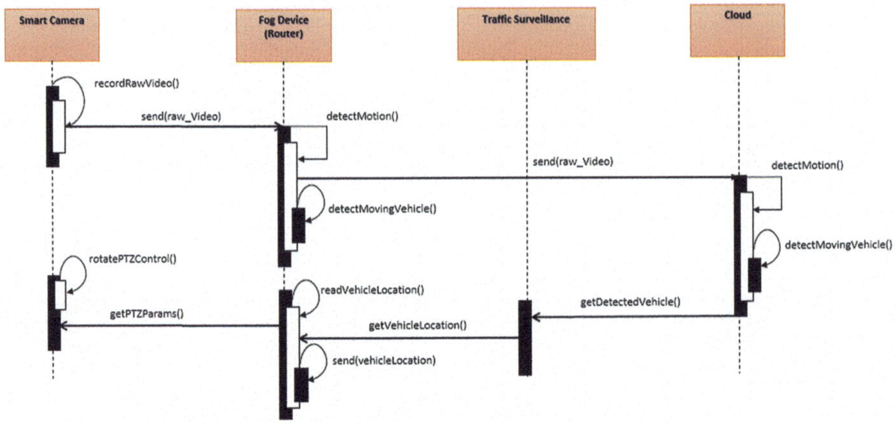

Fig. 3.3 UML sequence diagram of the proposed application model

can take place between objects, systems and subsystems. The figure also shows how messages are sent and processed using a vertical axis of the diagram. The Smart Camera records a raw video stream, and this is represented using *recordRawVideo()* method and then send the message using *send(rawVideo)* method which carries out this operation to the next object which fog device. Fog devices can be server, router, or controllers which are deployed throughout the network. When IoT sensors generate data, it is analysed via one of these nodes without having to be sent to the cloud. In this case, motion detection can be carried out in fog device using *detectMotion()* or a message can be sent to the cloud alternatively. Return messages are then sent back to the smart camera which acts as an IoT sensor and actuator of the system. These tasks are carried out in fog device and return vehicle location using *getVehicleLocation()* method and send PTZ params with *getPTZParams()* method. Alternatively, the *getDetectedVehicle()* method returns a message from the cloud to the Traffic Surveillance application when the current task was sent to the cloud, and to successfully carry out this task, a background image should be extracted from the original video image.

3.5.4 Activity Diagram of the Proposed Application Model

Figure 3.4 shows an activity diagram of our proposed model and how the tasks are being carried out for each instance of the Vehicle recognition.

The UML Activity Diagram demonstrated in Fig. 3.4 depicts the workflow of stepwise activities and the control flow of the proposed application model. It also defines how activities are coordinated to offer a service that can be at various levels of abstraction. Typically, each event of the Traffic Surveillance Application

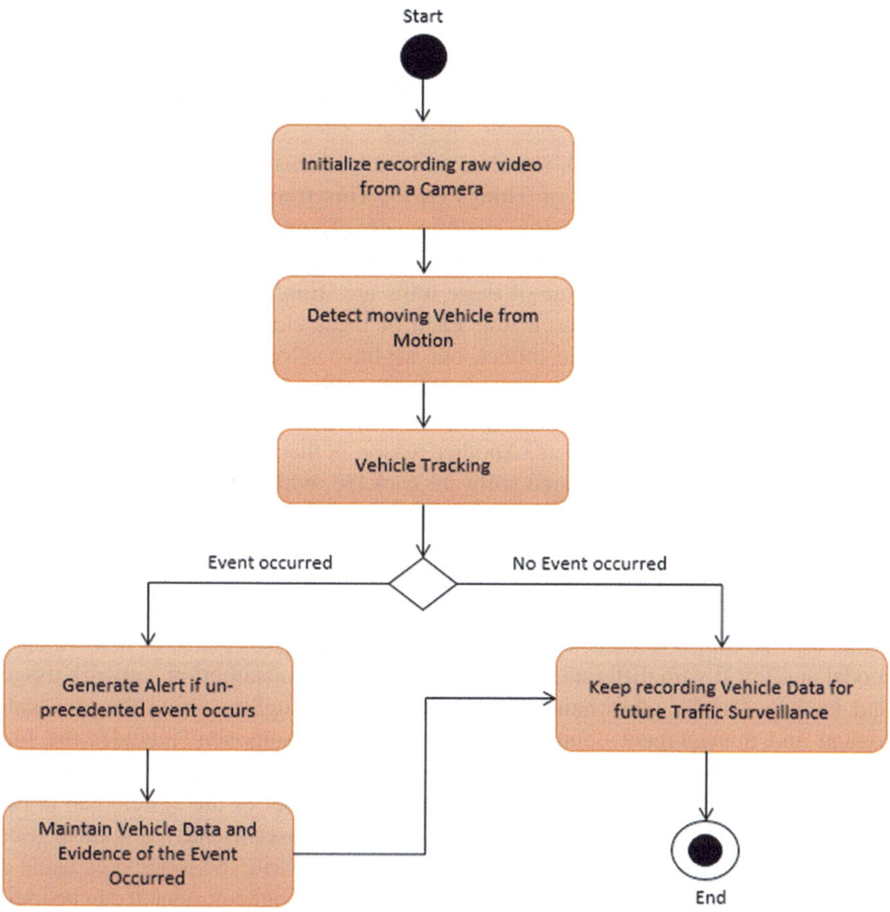

Fig. 3.4 UML activity diagram of proposed application model

needs to be accomplished by some operations, mainly where the operation needs to accomplish some different tasks. For the initial node which starts the beginning of a set of actions or activities, a black dot is used. The first object node of the activity initializes the recording of a raw video stream which is done using the Smart Video Camera. The following node represents the recognition of a vehicle from a video that contains a detected motion and then commences vehicle tracking by the last node. To control the flow of operation, a decision node is used to represent a test condition in a case where an unusual event is detected from a vehicle. If no event is occurred, a video camera continues to record video data for future surveillance, then the final node is used to stop all the control flow inactivity.

3.6 Performance Evaluation

3.6.1 Why Using iFogSim?

iFogSim is a toolkit to model, simulate and evaluate networks of fog computing, edge computing and Internet of Things (IoT). This framework offers a platform for analysing and evaluating the performance of applications. To model IoT applications, various simulator tools used in the fog computing environment are available in the literature. Some of these tools are: EmuFog [16], FogNetSim++ [17], FogTorchII [18], to name a few. Each of these simulation tools could have been used to test our proposed model, but we have selected iFogSim [19–21] for the following reason: iFogSim extends the CloudSim simulator by adding new functionalities [2], and as such, it is a preferred simulation toolkit for fog computing widely used by most researchers. Another reason is that iFogSim is written in the form of a Java-based open-source software package, which employs the JSON file format to represent the physical topologies, hence making it a package of choice to model real-time applications following the object-oriented paradigm. It also supports the simulation of entities and services. In iFogSim, the communication between entities/modules and the sharing of information between entities rely on a message passing mechanism. As such, iFogSim is a good platform to develop a model and test it with different metrics such as energy consumption, network usage and latency. On the other hand, the architecture of iFogSim offers the physical, logical and management components. The physical component includes the fog devices, the actuators and sensors, whereas the logical component includes the processing modules which are used to run tasks. The user of the system can draw, define the model, and build their topologies using a user-friendly graphical user interface (GUI). Alternatively, the user can define and write the topologies using a Java API which comes with the simulator. The management component can be used to schedule and monitor the application.

In our proposed model, we have chosen iFogSim because we have designed a tree-like topology that consists of a cloud data centre, gateways and smart cameras. We have designed the physical topology and defined it programmatically using Java APIs. The characteristics of each device and server have been defined using Java classes inherited from CloudSim.

3.6.2 Configuration Setup

The simulation of the Traffic Surveillance application was conducted using Window 10 Pro with Intel Core i5 processor, 8GB. The simulation was conducted for three sets of network topology configurations which are referred to as Configuration 1, Configuration 2 and Configuration 3, respectively. In Config 1, one area with one surveillance camera, Config 2, one area with two smart cameras, and lastly Config 3, with one area with three smart cameras. Each topology configuration was evaluated by using the four scheduling algorithms mentioned in Sect. 3.3, namely FCFS, SJF, GP and RR separately. The performance of these scheduling algorithms on vehicle tracking-task scheduling is evaluated based on average latency, energy consumption, execution time and network usage.

3.6.3 Simulation Results

Average Loop Delay

Figure 3.5 shows the performance of the studied four scheduling algorithms on the average latency for different configurations.

In Fig. 3.5, it can be observed that the latency increases as the number of areas surveilled also increases. In Config 1, all algorithms show a low latency below 100 milliseconds, and as the workload increases, the average latency also increases. This indicates that more fog devices are needed for effective processing, which in turn will minimize the time taken in the control loop to search for an efficient node. According to these results, one needs to add more fog devices to the network to minimize the latency as the workload increases. FCFS yields a very low latency in all three configurations when compared to other scheduling algorithms.

Energy Consumption

Figure 3.6 shows the performance of the studied scheduling algorithms on the energy consumption for different configurations. It is observed that as the number of smart cameras used in each configuration increases, the energy consumed also

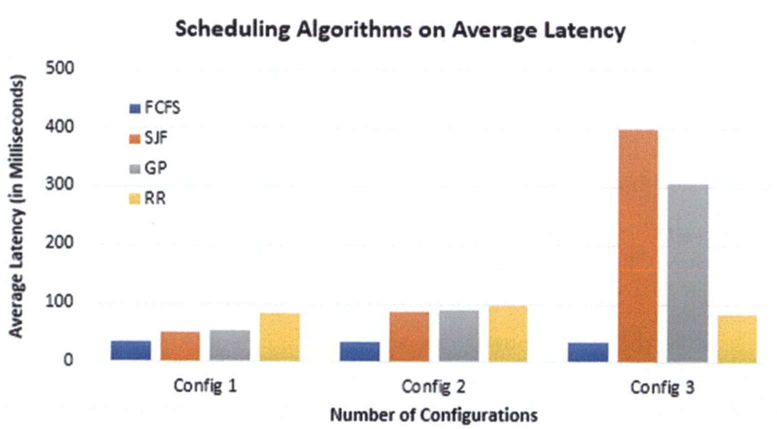

Fig. 3.5 Scheduling algorithms on average latency

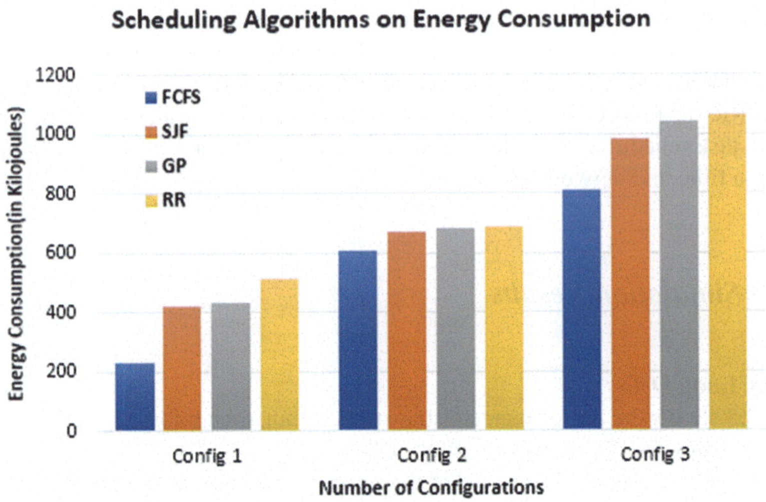

Fig. 3.6 Scheduling algorithms on energy consumption

increases. It also shows that to minimize the energy consumption in the network, fog devices must be added. Also, it is found that FCFS outperforms the other algorithms. In addition, RR appears to consume more energy while SJF and GP use almost equal energy in all setup configurations. These results also show that the best way to minimize the energy usage of our Traffic Surveillance application model is to execute the tasks in a queue and process them in the order of their arrivals.

Execution Time

Figure 3.7 shows the performance of the studied four scheduling algorithms on the execution time. It is observed that FCFS executes faster when compared to SJF and GP for the first two configurations. However, for Config 3, RR appears to outperform SJF and GP. It is also found that FCFS outperforms the other scheduling algorithms. RR appears to be less optimal in all configurations, this could be attributed to the fact that each ready task runs turn by turn only in a cyclic queue for a limited time, which may result in longer execution time compared to that obtained using other scheduling algorithms.

Network Usage

Figure 3.8 shows the performance of the studied four scheduling algorithms on network usage.

It is observed that as the number of surveillance cameras increases, several fog devices need to be added, which yields an increase in network usage. Also, the RR and GP algorithms appear to be superior to the other scheduling algorithms while SJF outperforms FCFS. Moreover, it is found that the tuple size and delay have a massive influence on network usage.

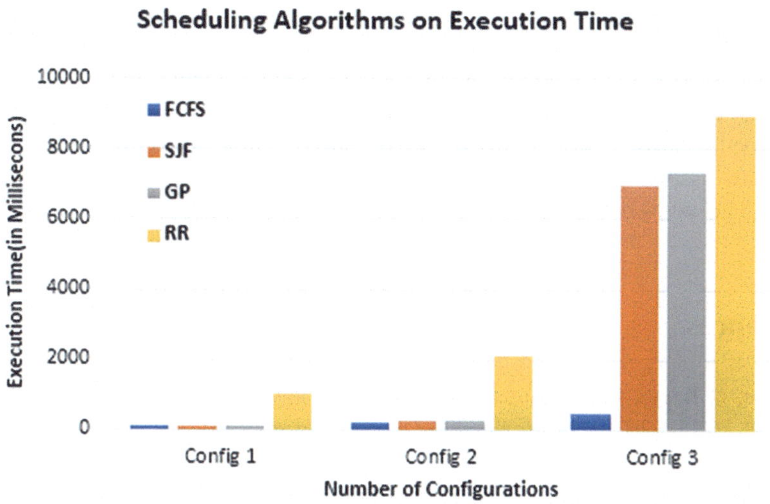

Fig. 3.7 Scheduling algorithms on execution time

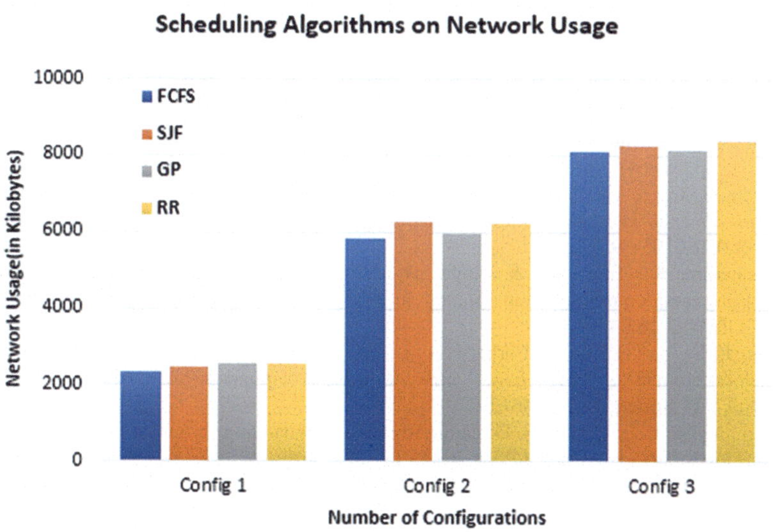

Fig. 3.8 Scheduling algorithms on network usage

3.7 Conclusion

In this chapter, a traffic surveillance model for detection and tracking of vehicles was proposed and simulations were conducted using iFogSim to evaluate its performance. The simulation results have shown that the FCFS algorithm outperforms the three other algorithms in terms of average latency, network usage, energy consumption and execution time, making it a suitable candidate for use in real-time IoT applications which require emergency response. In future work, we plan to simulate the proposed Traffic Surveillance application model in terms of resource usage and energy consumption in the scenario of multiple object tracking. We also plan to validate the effectiveness of our proposed Traffic Surveillance application model by assessing it against other available Traffic Surveillance application models, chosen as benchmark models.

References

1. Gubbi, J., Buyya, R., Marusic, S., & Palaniswami, M. (2013). Internet of things (IoT): A vision, architectural elements, and future directions. *Future Generation Computer Systems, 29*(7), 1645–1660. https://doi.org/10.1016/j.future.2013.01.010
2. Calheiros, R. N., Ranjan, R., Beloglazov, A., & De Rose, A. F. (2011). CloudSim: A toolkit for modeling and simulation of cloud computing environments and evaluation of resource provisioning algorithms. *Software: Practice and Experience, 41*, 23–50.
3. Hong, K., Lillethun, D., Ottenwälder, B., & Koldehofe, B. (2013). Mobile Fog: A programming model for large – Scale applications on the Internet of Things. In *MCC '13: Proceedings of the second ACM SIGCOMM workshop on mobile cloud computing* (pp. 15–20).
4. Chiu, C.-C., Ku, M.-Y., & Wang, C.-Y. (2010). Automatic traffic surveillance system for vision-based vehicle recognition and tracking. https://doi.org/10.6688/JISE.2010.26.2.17.
5. Sinqadu, M., & Shibeshi, Z. S. (2020). Performance evaluation of a traffic surveillance application using iFogSim. In *Lecture notes on data engineering and communications technologies* (Vol. 51, pp. 51–64). Springer Science and Business Media Deutschland GmbH.
6. Guevara, J. C., & Da Fonseca, N. L. S. (2021). Task scheduling in cloud-fog computing systems. *Peer-to-Peer Networking and Applications, 14*, 962–977. https://doi.org/10.1007/s12083-020-01051-9
7. Intharawijitr, K., Iida, K., & Koga, H. (2016). Analysis of fog model considering computing and communication latency in 5G cellular networks. https://doi.org/10.1109/PERCOMW.2016.7457059.
8. Oueis, J., Strinati, E. C., & Barbarossa, S. The fog balancing: Load distribution for small cell cloud computing. In *IEEE vehicular technology conference* (Vol. 2015). https://doi.org/10.1109/VTCSpring.2015.7146129
9. Aazam, M., & Huh, E. (2015). Dynamic resource provisioning through fog micro datacenter. https://doi.org/10.1109/PERCOMW.2015.7134002
10. Deng, R., Lu, R., Lai, C., & Luan, T. H. Towards power consumption-delay tradeoff by workload allocation in cloud-fog computing. In *IEEE international conference on communications* (pp. 3909–3914). https://doi.org/10.1109/ICC.2015.7248934
11. Pham, X. Q., & Huh, E. N. (2016). Towards task scheduling in a cloud-fog computing system. https://doi.org/10.1109/APNOMS.2016.7737240
12. Zeng, D., Gu, L., Guo, S., Cheng, Z., & Yu, S. (2016). Joint optimization of task scheduling and image placement in fog computing supported software-defined embedded system. *IEEE Transactions on Computers, 65*(12), 3702–3712. https://doi.org/10.1109/TC.2016.2536019
13. Gibet Tani, H., & El Amrani, C. (2018). *Smarter round robin scheduling algorithm for cloud computing and big data*. https://hal.archives-ouvertes.fr/hal-01443713
14. Mtshali, M., Africa, S., Adigun, M., Dlamini, S., & Mudali, P. Multi-objective optimization approach for task scheduling in fog computing. https://doi.org/10.1109/ICABCD.2019.8851038
15. Rahbari, D., & Nickray, M. Low-latency and energy-efficient scheduling in fog-based IoT applications. https://doi.org/10.3906/elk-1810-47

16. Mayer, R., Graser, L., Gupta, H., Saurez, E., & Ramachandran, U. *EmuFog: Extensible and scalable emulation of large-scale fog computing infrastructures*. https://github.com/emufog/emufog
17. Qayyum, T., Malik, A. W., Khattak, M. A. K., Khalid, O., & Khan, S. U. (2018). FogNetSim++: A toolkit for modeling and simulation of distributed fog environment. *IEEE Access, 6*, 63570–63583. https://doi.org/10.1109/ACCESS.2018.2877696
18. Brogi, A., Forti, S., & Ibrahim, A. (2017). How to best deploy your fog applications, probably. In *Proceedings - 2017 IEEE 1st international conference Fog Edge computing ICFEC 2017* (pp. 105–114). https://doi.org/10.1109/ICFEC.2017.8
19. Gupta, H., Vahid Dastjerdi, A., Ghosh, S. K., & Buyya, R. (2017). iFogSim: A toolkit for modeling and simulation of resource management techniques in the Internet of Things, Edge and Fog computing environments. *Software: Practice and Experience, 47*(9), 1275–1296. https://doi.org/10.1002/spe.2509
20. Naas, M. I., Boukhobza, J., Raipin Parvedy, P., & Lemarchand, L. (2018). An extension to iFogSim to enable the design of data placement strategies. In *2018 IEEE 2nd international conference on fog and edge computing, ICFEC 2018 - In conjunction with 18th IEEE/ACM international symposium on cluster, cloud and grid computing* (pp. 1–8). IEEE/ACM CCGrid. https://doi.org/10.1109/CFEC.2018.8358724
21. Mahmud, R., Narayana Srirama, S., Ramamohanarao, K., & Buyya, R. (2019). Quality of Experience (QoE)-aware placement of applications in Fog computing environments. *Journal of Parallel and Distributed Computing, 132*, 190–203. https://doi.org/10.1016/j.jpdc.2018.03.004

Chapter 4
Modifying the SMOTE and Safe-Level SMOTE Oversampling Method to Improve Performance

Kgaugelo Moses Dolo and Ernest Mnkandla

4.1 Introduction

The binary class distribution is a classification problem that involves two classes of the dataset. For a machine learning data model to generate accurate model outcome, the two classes of the dataset must be equally represented to avoid the issue of biasness [1–3]. Thus, oversampling and under-sampling are two techniques required to balance the class distribution of datasets in cases where the class distribution is skewed [1–4]. In this study, the original shape of the dataset represents the true context of the problem, which means that oversampling of the minority class and down-sampling of the majority class must be done to preserve the original shape of the dataset.

In general, the random selection of data observations from a population that has duplicates can result in a sample dataset that has duplicates. The implication is that, random sampling of a majority class that has duplicate data observations will result in duplicate data observations. It therefore means that, random sampling of majority class alone may not solve the problem of skewed class distribution for a binary classification problem. It is therefore suggested in this chapter that when down sampling the majority class, a random sampling on the majority class must be performed on a dataset that contains no duplicate data instances.

K. M. Dolo (✉) · E. Mnkandla
University of South Africa, Pretoria, South Africa
e-mail: 58527141@mylife.unisa.ac.za; mnkane@unisa.ac.za

I. Woungang, S. K. Dhurandher, *4th International Conference on Wireless, Intelligent and Distributed Environment for Communication*, Lecture Notes on Data Engineering and Communications Technologies 94, https://doi.org/10.1007/978-3-030-89776-5_4

For a highly skewed data distribution, re-oversampling of the minority class that negatively affects the original shape of the minority class data distribution must be avoided. Re-oversampling must be done to introduce new possibilities of the minority class rather than duplicating the existing minority class data instances. SMOTE is an oversampling method that is known to generate new instances of the minority class rather than duplicating the existing data instances of the minority class [5].

SMOTE oversamples the minority class by computing median feature vectors between a nominal feature sample and its potential nearest neighbours by Euclidean distance of standard deviations [5]. Duplicating minority training samples to reduce biasness of data distribution introduces high variance of data distribution. Thus, SMOTE is an important oversampling method that reduces variance of data distribution. The synthetic instances generated influence a classifier model to create less specific decision regions which are smaller.

Furthermore, positive influence can be better learned in general regions than positive instances subsumed around negative instances, which means that the SMOTE method suffers from a problem of generalization, whereby the region of a majority class is blindly generalized without considering the majority class [6, 7]. The generalization problem of the SMOTE method is particularly visible in a case of highly skewed class distribution since the minority class is thinly scattered in relation to the majority class. Thus, the probability of class mixture is very high. To keep the SMOTE method efficient and effective, an improvement of the algorithm is required.

Borderline SMOTE is an oversampling method designed to improve the performance of SMOTE oversampling method [1, 6]. The performance is improved by separating the positive instances into three regions namely borderline $\left(\frac{1}{2}K \leq n < K\right)$, noise ($n = k$) and safe $\left(0 \leq n < \frac{1}{2}K\right)$ [1, 6]. The three regions are separated while considering the negative instances on the K Nearest Neighbours. The borderline SMOTE uses the same oversampling method as SMOTE.

However, borderline SMOTE oversamples only the borderline instances of the minority class rather than oversampling entire instances of the minority class. Logically, the two consecutive instances are obviously not different, but they are divided into two regions (noise and borderline), whereby the first instance is selected for oversampling and the other instance is declined for oversampling.

Another method is Safe-Level SMOTE, which is an oversampling method that creates safe-level synthesis of the minority class [1, 6, 7]. The synthetic instances are placed closer to the safe level; that is to say, the safe level closer to K is nearly noise. The safe level is defined as the number of positive instances within K Nearest Neighbour but not equal to K Nearest Neighbour [6]. The Safe-Level SMOTE is therefore found to be a promising algorithm with a positive impact in this study.

The rest of the chapter is organized as follows. Section 4.2 presents the research questions addressed by this work and their objectives. Section 4.3 presents the Modified SMOTE method. Section 4.4 presents the simulation results, and Sect. 4.5 concludes the chapter.

4.2 Research Questions and Objectives

4.2.1 Research Questions

Since the banking sector is migrating from data analytics to data insights in South Africa [8], the study therefore sets out to explore the following main research question: Will the oversampling of the minority class that generates new data instances at the safe level of the minority class and under-sampling of the majority class that randomly selects non-duplicate data instances preserve the nature of the dataset and the original context of the problem for credit card fraud data when dealing with highly skewed class distributions? This main research question is broken down into the following sub-questions:

(a) Does Safe-Level SMOTE oversampling method (on minority classes) used with under-sampling method (that eliminates duplicate data samples on majority class) have positive impact on reducing the high skewedness of the class distribution than SMOTE oversampling method (on minority classes) used with under-sampling method (that also eliminates duplicate data samples on majority class)?
(b) Do Safe-Level-SMOTE oversampling method and the under-sampling method (that eliminates duplicate data samples on majority class) reduce or eliminate the problem of overlapping data samples between fraudulent and non-fraudulent classes?

This study therefore sought to answer these questions by carrying out the objectives presented in the next section.

4.2.2 Research Objectives

The main objective of this study is to obtain a balanced class distribution of credit card fraud data that preserves the nature of the dataset and the original context of the problem. The main objective is further broken down into two sub-objectives presented below:

(a) To determine if modified Safe-Level SMOTE oversampling method used with under-sampling method has a positive impact on reducing the high skewedness of the class distribution than the modified SMOTE oversampling method used with under-sampling method. This objective will be implemented by developing and running a modified SMOTE algorithm and a modified Safe-Level SMOTE algorithm on the minority class data, and an under-sampling algorithm on the majority class data.
(b) To investigate if the modified Safe-Level SMOTE oversampling method and the under-sampling method reduce or eliminate the problem of overlapping data samples between fraudulent and non-fraudulent classes. This objective will be

implemented by running a modified Safe-Level SMOTE on minority class data and an under-sampling algorithm on majority class data.

The next section discusses the research design followed to achieve the research objectives of this study.

4.3 Research Design

The modified SMOTE method in this study will be compared to the SMOTE method, and the modified Safe-Level SMOTE method will be compared to the Safe-Level SMOTE. The modified SMOTE and the modified Safe-Level SMOTE were used to oversample the data observations of the minority class of the dataset together with down-sampling method that removes duplicates and randomly chooses non-duplicate data samples from the majority class in order to control the best fit line to an optimum place that represents fraudulent transactions and non-fraudulent transactions equally. A dataset with equal representation of fraudulent transactions and non-fraudulent transactions makes it easier for a classification model to learn data of the two classes well. Thus, the modified SMOTE and modified Safe-Level SMOTE were tested by running Artificial Neural Network, Support Vector Machine, Naïve Bayesian and k-Nearest Neighbours algorithms.

While the objective of the SMOTE algorithm is to generate a new data instance between two existing data instances of the minority class [5]. This chapter argues that the logic used to generate new data instances by the SMOTE does not always generate a new data instance between two given data instances.

From the **for loop** of attributes to the above SMOTE Algorithm 1, the algorithm is looping through the attributes of the dataset. The synthetic instance is generated by finding the difference of attribute values between the data instance of interest and its chosen nearest neighbour. The difference is then multiplied by the gap which is the random value chosen between 0 and 1. The multiplication of the difference and the gap is added to the data instance of interest.

Although this method works well, there are limitations to its objective. The limitations are introduced by the logic behind the mathematics of the discussed attribute loop of the SMOTE method. This chapter therefore suggests that if two data instances have values of opposite signs, then the SMOTE algorithm will not generate a new data instance that is between the two data instances, which violates the objective of SMOTE algorithm. The difference in attribute loop of SMOTE will change to addition, given that the two data instances have values of opposite signs.

Below is a demonstration of the above claim using the attribute loop of SMOTE. The attribute loop of SMOTE is where the new data instance is generated between two data points. Figure 4.1 shows the function of attribute loop of SMOTE.

The following random values of instances will be chosen to test the claim made in this study: instances = [[−10, −21, −4, −45, −66, −93, 1, 10, 21, 4, 45, 66, 93, 1]]. To pictorially see the output, the x-axis values are required. The standard

```
Algorithm 1: SMOTE
Input: Minority data observations-T, Size of new
instance in percentage-N,
      Size of the nearest neighbours-K
  Output: I=(N/100)*count(T) - is the newly generated
minority class instances of
                   size I - Synthetic-instances.
  Begin
   1. No-of-minority = count(T)
   2. No-new-Instances = I
   3. No-of-attributes = count(T[0] - 1
   4. Minority-instances = T
   5. Synthetic-instances= []
   6.
   7. For positive-instance in positive-instances:
   8.       Compute 5 neighbours of positive-
      instance and call it nnarray.
   9.       For index in No-new-instances:
   10.                Choose a random number
      between 0 and number of neighbours and call it
      nn.
   11.                For attribute in attributes:
   12.                      difference =
      nnarray[nn][attribute] -positive-
      instance[attribute].
   13.                      Gap = generate random
      value between 0 and 1.
   14.                      Synthetic-
      instances.append(positive-instance[attribute]+
      difference * gap)
   15.     Append minority class value to Synthetic-
      instance
End of the function
```

numbering system from 1 to 7 will be used for visualization and for demonstration purposes. Figure 4.2 is the pictorial representation of the claim.

Figure 4.2 shows that the SMOTE instance is generated outside the boundaries of data instance 1 and data instance 2, which means that the claim made in this study about SMOTE method is correct. The larger the space between the value of data instance 1 and data instance 2, the larger the space between the generated data value and the value of data instance 1 and data instance 2. To solve this problem, the difference in the attribute loop of the SMOTE method must be modified to handle the issue of opposite signs.

```
def While_SMT(instances):
    Sythetic_values = []
    for No_New_instances in range(len(instances)):
        row = []
        for attr in range(len(instances[0]) - 1):
            difference = float(instances[0][attr]) - float(instances[1][attr])
            gap = np.random.uniform(0, 1)
            row.append(instances[No_New_instances][attr] + (difference * gap))
        row.append(int(1))
        Sythetic_values.append(row)
    return Sythetic_values
```

Fig. 4.1 SMOTE function

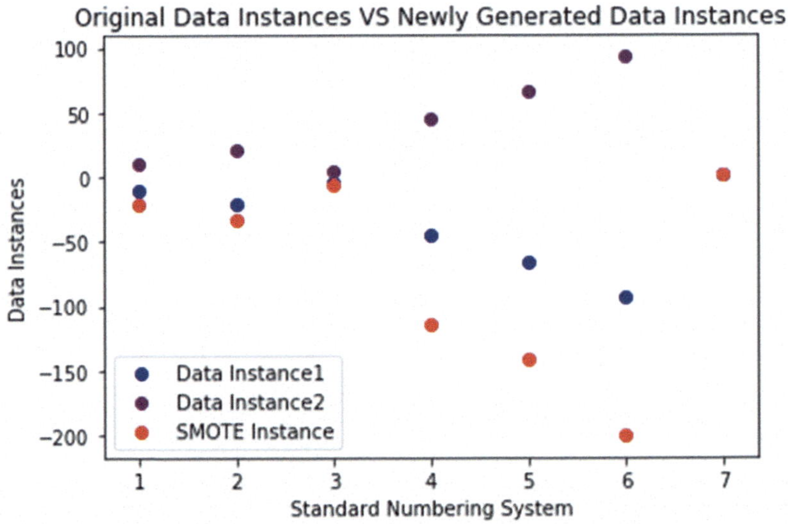

Fig. 4.2 SMOTE demonstration

In this chapter, we propose a random generation of values between the two values of data instance 1 and data instance 2 as a way to ensure that the newly generated values lie within the boundary of data instance 1 and data instance 2. Figure 4.3 shows the modified function of the attribute loop of SMOTE.

The function will use the same values as the above SMOTE function to maintain consistence. Figure 4.4 is the pictorial representation of the claim made by this study that newly generated random values will lie within the boundary of data instance 1 and data instance 2 regardless of the signs.

Figure 4.4 shows that the SMOTE instance is generated within the boundaries of data instance 1 and data instance 2, which means that the claim made in this study about SMOTE method is absolutely correct. Thus, below is the modified algorithm

```
def While_SMT_New(instances):
  Sythetic_values = []
  for No_New_instances in range(len(instances)):
    row = []
    for attr in range(len(instances[0]) - 1):
      row.append(np.random.uniform(instances[0][attr], instances[1][attr]))
    row.append(int(1))
    Sythetic_values.append(row)
  return Sythetic_values
```

Fig. 4.3 SMOTE function (modified)

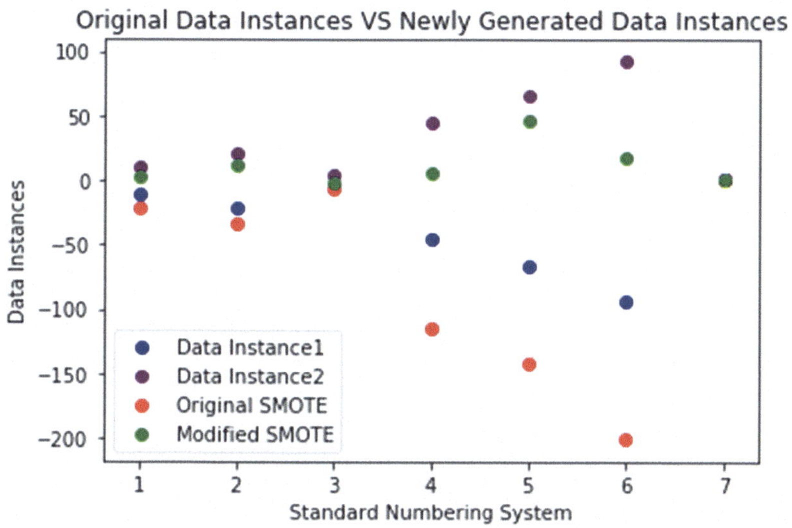

Fig. 4.4 Modified SMOTE demonstration

of SMOTE method that ensures that newly generated data values are generated within the boundaries of data instance 1 and data instance 2.

By generating data instances within the boundaries of data instance 1 and data instance 2, we achieve the objective of the original SMOTE method. Thus, the comparison of SMOTE and safe-level SMOTE algorithms has merit. Below is the definition of the algorithm of safe-level SMOTE.

From the above algorithm, it can be seen that the new data instances that are generated follow the same logic as the SMOTE method above except the fact that the newly generated data values in this case are generated at the safe level of SMOTE. Again, to preserve the primary objective of the SMOTE algorithm, the new data values of safe-level SMOTE must be generated between the data instances at the safe-level of SMOTE. Below is the modified safe-level SMOTE algorithm.

```
Algorithm 2: Modified SMOTE
Input: Minority data observations-T, Size of new
instance in percentage-N,
     Size of the nearest neighbours-K
  Output: I=(N/100)*count(T) - is the newly generated
minority class instances of
                    size I - Synthetic-instances.
  Begin
   1. No-of-minority = count(T)
   2. No-new-Instances = I
   3. No-of-attributes = count(T[0] - 1
   4. Minority-instances = T
   5. Synthetic-instances= []
   6.
   7. For positive-instance in positive-instances:
   8.       Compute 5 neighbours of positive-
      instance and call it nnarray.
   9.       For index in No-new-instances:
   10.                Choose a random number
      between 0 and number of neighbours and call it
      nn.
   11.                For attribute in attributes:
   12.
      row.append(np.random.uniform(instances[0][attr
      ], instances[1][attr])
   13.      Append minority class value to Synthetic-
      instance
  End of the function
```

The demonstration of how new values of data instances are generated is done from the SMOTE algorithm above. In the data analysis section below, the study compares the performance of the SMOTE and the safe-level SMOTE algorithms.

4.4 Simulation Results

In the data analysis section, the study compares the performance of SMOTE and safe-level SMOTE algorithms, and the modified SMOTE and safe-level SMOTE algorithms by supplying machine learning algorithms with a binary classification dataset that is down sampled by removing duplicate data samples and randomly choose the data observations. The machine learning algorithms chosen in this study

```
Algorithm 3: Safe Level SMOTE
Input: Minority data observations-T, Size of new
instance in percentage-N,
     Size of the nearest neighbours-K
  Output: I=(N/100)*count(T) - is the newly generated
minority class instances of
                  size I - Synthetic-instances.
  Begin
   1. No-of-minority = count(T)
   2. No-new-Instances = I
   3. No-of-attributes = count(T[0] - 1
   4. Minority-instances = T
   5. Synthetic-instances= []
   6. For each positive-instance in positive-
      instances:
   7.     Compute 5 neighbours of positive-instance
      and call it K.
   8.     Compute 5 neighbours of positive-instance
      and call it N.
   9.     Count number of positive instances in K
      and call it SLp.
   10.         Count number of positive instances in
      N and call it SLn.
   11.         If SLp != 0:
   12.             SL-ratio = SLp/SLn
   13.         Else:
   14.             SL-ratio = np.inf
   15.         If SL-ratio == np.inf AND SLp ==0:
   16.             Continue.
   17.         Else:
   18.             Feature = []
   19.             For index in
      range(len(np.array(T)[0])):
   20.                 If SL-ratio == np.inf AND
      SLp != 0:
   21.                     Gap = 0
   22.                 Elseif SL-ratio == 1:
   23.                     Gap =
      np.random.uniform(0,1)
   24.                 Elseif SL-ratio > 1:
   25.                     Gap =
      np.random.uniform(0,1 / SL-ratio)

   26.             Elseif SL-ratio < 1:
   27.                 Gap = np.random.uniform(1 -
      SL-ratio , 1)
   28.             difference = nnarray[nn][attribute]
      -  positive instance[attribute].
   29.             Gap = generate random value between
      0 and 1.
   30.             Synthetic-instances.append(positive
      - instance[attribute]+ difference * gap)
   31. Append minority class value to Synthetic-
      instance.
End of the function
```

```
  Algorithm 4: Modified Safe Level SMOTE
Input: Minority data observations-T, Size of new
instance in percentage-N,
     Size of the nearest neighbours-K
  Output: I=(N/100)*count(T) - is the newly generated
minority class instances of
                size I - Synthetic-instances.
  Begin
   1. No-of-minority = count(T)
   2. No-new-Instances = I
   3. No-of-attributes = count(T[0] - 1
   4. Minority-instances = T
   5. Synthetic-instances= []
   6. For positive-instance in positive-instances:
   7.     Compute 5 neighbours of positive-instance
      and call it K.
   8.     Compute 5 neighbours of positive-instance
      and call it N.
   9.     Count number of positive instances in K
      and call it SLp.
   10.         Count number of positive instances in
      N and call it SLn.
   11.         If SLp != 0:
   12.             SL-ratio = SLp/SLn
   13.         Else:
   14.             SL-ratio = np.inf
   15.         If SL-ratio == np.inf AND SLp ==0:
   16.             Continue.
   17.         Else:
   18.             Attribute = []
   19.             For index in
      range(len(np.array(T)[0])):
   20.                  If SL-ratio == np.inf AND
      SLp != 0:
   21.                        Gap = 0
   22.                  Elseif SL-ratio == 1:
   23.                        Gap =
      np.random.uniform(0,1)
   24.                  Elseif SL-ratio > 1:
   25.                        Gap =
      np.random.uniform(0,1 / SL-ratio)
   26.                  Elseif SL-ratio < 1:
   27.                        Gap =
      np.random.uniform(1 - SL-ratio , 1)
   28.
      Attribute.append(np.random.uniform(nnarray[nn]
      [attribute],positiveinstances[attribute])
   29.                  Attribute.append[-1] = 1.0
   30.                  Synthetic-
      instance.append(Attribute)
   31.      Append minority class value to
      Synthetic-instance.
  End of the function
```

Table 4.1 Performance comparison of the studied algorithms

	SMOTE	SL SMOTE	SMOTE modified	SL SMOTE modified
Artificial neural network	99.58	99.79	99.58	99.37
Support vector machine	96.46	97.27	96.46	96.86
Naïve Bayesian	100.00	100.00	100.00	100.00
Decision tree	100.00	100.00	100.00	100.00
K-nearest neighbour	97.08	95.38	97.08	94.55

Table 4.2 Confusion matrix: true positives

	SMOTE	SL SMOTE	SMOTE modified	SL SMOTE modified
Artificial neural network	74	66	72	69
Support vector machine	73	64	69	65
Naïve Bayesian	72	67	72	70
Decision tree	74	67	72	70
K-nearest neighbour	69	54	70	70

are Artificial Neural Network, Support Vector Machine, Naïve Bayesian and k-Nearest Neighbour. Table 4.1 shows the performance of SMOTE algorithms in relation to the chosen machine learning algorithms.

The results demonstrate that the oversampling of the minority class plays an important role as shown by the high accuracy of all machine learning models. The oversampling of the minority class was done by generating new instances of the dataset given majority data observations that are down sampled by removing duplicate data samples and randomly choosing the data observations.

The objective of this chapter was to modify the SMOTE and safe-level SMOTE algorithms and compare the performance of original algorithms and modified algorithms. These results show that the safe-level SMOTE algorithm performs better than SMOTE algorithms on credit card fraud data. Furthermore, these results show that the modified SMOTE algorithm performs better than the modified safe-level SMOTE algorithm.

Evaluate Model

The confusion matrix is an evaluation technique that has been used to evaluate the performance of the classification model. The confusion matrix is used to compute the true-positive, false-positive, false-negative and true-negative transactions. The smaller the percentage of false positives and false negatives indicate that the model is actually performing well. Tables 4.2, 4.3, 4.4, and 4.5 show the output of the confusion matrix evaluation technique for each classification model.

The false-negative and false-positive tables show output values which are very small. These small output values indicate that the classification models against oversampling techniques are performing very well.

Table 4.3 Confusion matrix: false positives

	SMOTE	SL SMOTE	SMOTE modified	SL SMOTE modified
Artificial neural network	0	3	0	2
Support vector machine	6	2	2	0
Naïve Bayesian	0	0	0	0
Decision tree	0	0	0	0
K-nearest neighbour	7	12	2	7

Table 4.4 Confusion matrix: false negatives

	SMOTE	SL SMOTE	SMOTE modified	SL SMOTE modified
Artificial neural network	0	1	0	1
Support vector machine	1	3	3	5
Naïve Bayesian	0	0	0	0
Decision tree	0	0	0	0
K-nearest neighbour	5	13	7	2

Table 4.5 Confusion matrix: true negatives

	SMOTE	SL SMOTE	SMOTE modified	SL SMOTE modified
Artificial neural network	381	381	383	378
Support vector machine	375	382	381	380
Naïve Bayesian	381	384	383	380
Decision tree	381	384	383	380
K-nearest neighbour	374	372	376	376

4.5 Results

In conclusion, it can be said that both the SMOTE and safe-level SMOTE are powerful oversampling techniques when they are used with majority data observations that are down sampled by removing duplicate data samples and randomly choosing the data observations. Given that SMOTE and safe-level SMOTE algorithms were modified in this study and the output shows that modified SMOTE performs better than modified safe-level SMOTE; this study therefore concludes that incorrect computation of an algorithm can cloud the algorithm's true computing capabilities.

Acknowledgments I would like thank Prof Ernest Mnkandla for providing insight, expertise and comments that greatly assisted the research. I would also like thank my family: Joyce Noko Dolo, Mahlatse Dolo, Kamogelo Dolo, Tshegofatso Kgoele, Flora Mahlodi Dolo, etc. for their support.

References

1. Gosain, A., & Sardana, S. (2017). Handling class imbalance problem using oversampling techniques: A review. In *2017 international conference on advances in computing, communications and informatics (ICACCI)* (pp. 79–85).
2. Matsuda, K., & Murase, K. (2016). Single-layered complex-valued neural network with SMOTE for imbalanced data classification. In *2016 joint 8th international conference on soft computing and intelligent systems (SCIS) and 17th international symposium on advanced intelligent systems (ISIS)* (pp. 349–354).
3. Pengfei, J., Chunkai, Z., & Zhenyu, H. (2014). A new sampling approach for classification of imbalanced data sets with high density. In *2014 international conference on big data and smart computing (BIGCOMP)* (pp. 217–222).
4. Zeager, M., et al. (2017). Adversarial learning in credit card fraud detection. In *Systems and information engineering design symposium (SIEDS)* (pp. 112–116).
5. Chawla, N. V., Bowyer, K. W., Hall, L. O., & Kegelmeyer, W. P. (2002). SMOTE: Synthetic minority over-sampling technique. *Journal of Artificial Intelligence Research, 16*, 321–357.
6. Bunkhumpornpat, C., Sinapiromsaran, K., & Lursinsap, C. (2009). Safe-level-SMOTE: Safe-level-synthetic minority over-sampling technique for handling the class imbalanced problem. In *Pacific-Asia conference on knowledge discovery and data mining* (pp. 475–482).
7. Meidianingsih, Q., Erfiani, & Sartono, B. (2017). Study of safe-level SMOTE method in unbalanced data classification cases. *International Journal of Scientific and Engineering Research, 8*(5), 1167–1171.
8. TCI. (2015). Driving profitable growth through improved data insights and segmentation practices for the future customer and digital bank. In *Indaba Hotel–Sandton, trade conferences international* (pp. 1–7).

Chapter 5
Performance Evaluation of Fuzzy-Based Routing Protocols for Opportunistic Networks

Khuram Khalid, Isaac Woungang, Sanjay K. Dhurandher, and Jagdeep Singh

5.1 Introduction

Opportunistic networks (OppNets) are characterized by periodic connectivity and high node mobility [1]. In this context, the message in such networks is carried progressively from source to destination using the store-carry-and-forward mechanism [2] whereby a node that receives the data packets stores them in its local buffer (if space is available), awaiting for a forwarding opportunity to pass these packets to a suitable one-hop node qualified to carry them towards the destination. If no space is available in the node's buffer, the message is considered as lost in the network.

Various routing protocols for OppNets implementing the aforementioned store-carry-and-forward strategy [2] in conjunction with other techniques have been investigated in the literature. Representative ones are reported in [3–13]. This chapter focuses on fuzzy-based routing schemes for OppNets. Precisely, it reports on additional simulation studies of three recently proposed fuzzy-based routing protocols under both synthetic and real mobility traces, namely the RLFGRP [3], FCSG [4], and MARL-CC [5] schemes, in terms of delivery ratio, overhead ratio,

K. Khalid (✉) · I. Woungang
Department of Computer Science, Ryerson University, Toronto, ON, Canada
e-mail: khuram.khalid@ryerson.ca; iwoungan@ryerson.ca

S. K. Dhurandher
Department of Information Technology, Netaji Subhas University of Technology, New Delhi, Delhi, India

J. Singh
Department of Computer Science and Engineering, Sant Longowal Institute of Engineering and Technology, Longowal, Punjab, India
e-mail: Jagdeep@sliet.ac.in

I. Woungang, S. K. Dhurandher, *4th International Conference on Wireless, Intelligent and Distributed Environment for Communication*, Lecture Notes on Data Engineering and Communications Technologies 94, https://doi.org/10.1007/978-3-030-89776-5_5

and average latency, under varying number of nodes, time to live (TTL), and buffer size.

The remainder of the chapter is organized as follows. Section 5.2 presents an overview of the studied protocols. In Sect. 5.3, a performance comparison of the studied protocols is presented. Section 5.4 concludes the chapter.

5.2 Overview of Protocols

In this section, an overview of the studied routing protocols for OppNets is provided.

5.2.1 RLFGRP Protocol

The following notations are used in the FCSG algorithm:

- SN: Sender node
- RN: Receiver node
- MFNL: Message Forwarding Nodes List (Hashtable)
- m: Message
- MFS: Message Forwarding Set

The design of the Reinforcement Learning-based Fuzzy Geocast Routing Protocol (RLFGRP) [3] is illustrated in Algorithm 1. The scheme consists of two phases. In the first phase, a Q-learning mechanism is applied for the forwarding of the message towards the destination cast. More precisely, when a source node, say S, creates a message m to be sent to a destination node, say D, the destination geocast region is instantly defined, along with the associated cast definition, which represents an ensemble of two-dimensional points forming the geographic cast and a pair of epoch times defining the message lifetime.

Whenever a forwarding opportunity arises, i.e., the source node S encounters an intermediate node, say N, the likelihood of N to carry the message towards the destination cast is calculated by means of a fuzzy controller which takes the Q-value, reward value, and remaining buffer space of N as input parameters. Typically, the considered Q-learning mechanism is a continuous process out of which a reward (or penalty in the form of reduced reward) is assigned to node N (according to a reward function), along with its Q-value, based on the action it has taken. It should be noted that this reward function takes the Euclidean distance from node N to the source node S, the delivery probability of N, and the direction of N with respect to destination node D, as input parameters. Besides, during this Q-learning iterative process, the update of the Q-value of a node is done on the basis of its assigned reward value, its action, and the algorithm's discount and learning rates. Once the likelihood of any intermediate encountered node is calculated, it is verified whether that node belongs or not to a prescribed Message Forwarding Set (MFS), and if that

Algorithm 1 RLFGRP routing scheme [3]

```
Initialization during message generation:
m is created
Sender node and destination geocast region are defined

Phase 1: Moving the message towards its destination cast
for each SN that meets the next RN do
    Drop the expired messages from buffer
    for each m in the SN buffer do
        if m already exists in RN then
            skip and go to the next m in the for loop
        else if If m is not in the geocast region then
            for each RN in the neighbours of SN do
                Calculate the likelihood of RN
                if likelihood is from MFS then
                    Store RN in MFNL List
                end if
            end for
            if MNFL is empty then
                skip and go to the next m in the for loop
            end if
            for each node J in MFNL do
                Update the Q-Value of RN
            end for
            if Q-value of RN is > than J's q-value then
                Forward message to the current RN
                Update the reward of the current RN
            end if
Phase 2: Distributing the message copy to all nodes within the geocast region
        else if m is in geocast region then
            if RN already exists in geocast region then
                if m does not exist in RN then
                    Forward a copy of m to RN
                end if
            end if
        end if
    end for
end for
```

is the case, that node's ID is referenced in a hash table and the Q-values of all nodes in this table are updated, then the node in this list which has the highest Q-value is selected as best forwarder of the message m towards its destination. In the second phase, a Check-and-Spray mechanism similar to that used in [13] is implemented to intelligently flood the message within the geocast region, hoping that it will reach its intended destination.

5.2.2 FCSG Protocol

The following notations are used in the FCSG algorithm:

- C: Number of remaining replicas of a message
- VH: Very high
- H: High
- M: Medium
- L: Low
- VL: Very low
- SN: Sender node
- RN: Receiver node
- m: Message
- L(): Likelihood function

The design of the Fuzzy-based Check-and-Spray Geocast Routing Protocol (FCSGRP) for opportunistic networks [4] is illustrated in Algorithm 2. The scheme consists of two phases. In the first phase, a multi-copying spray mechanism is applied for the forwarding of the message towards the destination cast. More precisely, similar to the previous RLFGRP scheme, when a source node, say S, creates a message m to be sent to a destination node, say D, it is initially allocated a payload data. The destination geocast region is also instantly defined, along with the cast definition, and C an integer value representing the maximum number of remaining message copies that can be generated by S and forwarded eventual encountered nodes during the whole message forwarding process is set.

Whenever a forwarding opportunity arises, i.e., the source node S opportunistically meets an intermediate node, the likelihood of this node to carry the message towards the destination cast is calculated by using the series of two fuzzy controllers. In this process, the first controller considers as inputs the speed and direction of that intermediate node to determine its movement, noting that the direction here is defined as the angle between the line joining node S to the centre of the geocast region and the line on which this intermediate node currently moves. The second controller uses the movement, remaining energy, and buffer space of that intermediate node as inputs to calculate its likelihood to carry the message towards the destination cast. Once the likelihood of any intermediate encountered node, say N, is calculated, it is verified whether this value is greater than the likelihood of the source node S. If that is the case, the message m is forwarded to node N and its C value gets decreased based on the likelihood fuzzy controller's output; otherwise, node N is deemed as not qualified to receive the message and the value of C is kept unchanged, and the search for a newly suitable intermediate node to carry the message towards the destination cast recommences. This process stands as long as the value of C is not equal to 0. When this value becomes 0 and the message

Algorithm 2 FCSG routing scheme [4]

```
Initialization during message generation:
m is created
Sender node and destination geocast region are defined
C is initialized

Phase 1: Moving a message towards its destination cast
for each SN that meets the next RN do
    drop expired messages from buffer
    for each m in the SN buffer do
        if m already exists in RN then
            skip this m and go to the next m in the for loop
        end if
        if C > 0 then
            if RN is located in geocast area then
                forward message to RN
                C = 0
                // In case RN is not located within the
                // geocast region, calculate the likelihood of
                // RN to receive the message.
            else if L(RN,m) > L(SN,m) then
                forward a copy of message to RN
                switch L(RN,m) do
                    case VH: C = C-5, break;
                    case H: C = C-4, break;
                    case M : C = C-3, break;
                    case L : C = C-2, break;
                    case VL : C = C-1, break;
            end if
Phase 2: Distributing message copy to all nodes inside the geocast
        else if C = 0 and m is in geocast region then
            if RN already exists in geocast region then
                if m does not already exist in RN then
                    forward a copy of m to RN
                end if
            end if
        end if
    end for
end for
```

has not yet reached the destination cast, no more message copies are generated and the message is considered as lost in the network. In the second phase, the same Check-and-Spray mechanism described in [13] is used as controlled flooding technique within the geocast region, hoping that the message will eventually get to its destination.

5.2.3 *MARL-CC Protocol*

The design of the Multi-Agent Reinforcement Learning Congestion Control (MARL-CC) [5] involves the use multiple agents present in the network environment to perform the routing of the message. As such, the messages in the network are considered as the agents and the nodes that participate in the message routing from source to destination are considered as the states. The protocol works as follows.

Whenever a message (agent) originated from a source node arrives at a given node (i.e., intermediate node), the Q-table's row for the address of the destination node is searched up and a global Q-table representing the set of all tables to be used in the routing process is built and learned using the so-called Train Routers algorithm. In this learning process, the Q-tables are meant to guide the agent to the next best forwarder nodes to carry it to its destination in the sense that the agent is most likely directed towards a routing path that achieves the maximum reward globally based on its action's effectiveness in controlling the network congestion level. Next, once the Q-tables for all nodes are populated, a Q-learning algorithm is invoked which considers the set of IDs of the neighbouring nodes of a node, the message to be forwarded, and the message destination, as input parameters to calculate (and update) the Q-value of nodes. Based on this, intelligent routing decisions regarding the selection of candidate forwarders for the message are determined by means of an implemented congestion controlled optimal policy that makes these routing decisions also contribute to the improvement of the message delivery probability in the sense that the network overhead is shown through simulations to be significantly reduced. The pseudocode of the MARL-CC algorithm is given in [5].

5.3 Performance Evaluation

In this section, the studied routing protocols for OppNets, namely RLFGRP [3], FCSG [4], and MARL-CC [5], are simulated and compared in terms of delivery ratio, average latency, and overhead ratio, under varying number of hosts, buffer size, and time to live (TTL). In this work, the delivery ratio is calculated as the ratio of the messages successfully delivered to the destination at the end of the simulations and the total number of messages generated in the network. The overhead ratio is a measure of the bandwidth efficiency, calculated as

$$\frac{\text{Number of relayed messages} - \text{Number of delivered messages}}{\text{Number of delivered messages}} \tag{5.1}$$

Algorithm 3 MARL-CC Routing

```
Input:
msgList[1..numNodes] = null,
neighbors[1..numNodes] = null,
neighborsPrev[1..numNodes] = null,
destNode[1..numMessages],
globalQ[nodeA][destNode[message]][nodeB],globalQ[nodeB][destNode[message]][nodeA],
nodeA, nodeB
Procedure
for each message in msgList[nodeA] do
    if globalQ[nodeA][destNode[message]][nodeB] is updated then
        Calculate reward using congestion factor CF_x
        select nextHop by Boltzmann Exploration scheme on
        the set {(x, globalQ[nodeA][destNode[message]][x])
        textbar x ∈ neighbors[nodeA]}
        forwardMessage(message, nodeA, nextHop)
    else
        Update globalQ[nodeA][destNode[message]][nodeB]
    end if
end for
for each message in msgList[nodeB] do
    if globalQ[nodeB][destNode[message]][nodeA] is updated then
        Calculate reward using congestion factor CF_x
        select nextHop by Boltzmann Exploration scheme on the
        set {(x, globalQ[nodeB][destNode[message]][x])
        |x ∈ neighbors[nodeB]}
        forwardMessage(message, nodeB, nextHop)
    else
        Update globalQ[nodeB][destNode[message]][nodeA]
    end if
end for
update (msgList[nodeA], msgList[nodeB])
update (neighbors[nodeA], neighbors[nodeB]) and (neighborsPrev[nodeA],
neighborsPrev[nodeB])
return
```

Two mobility models are considered: (1) the Shortest Path Map-Based Movement (SPMBM) whose system model is a set of six groups of nodes, i.e., electric motor cars, pedestrian, bicycles, tram, vehicles, and office workers, that can leave or join the network at any time, and these nodes move on a shortest path determined by the Dijkstra algorithm, and (2) the INFOCOM 2006 dataset of real mobility traces [14]. The wireless interfaces used are Wi-Fi 802.11ac with a transmission speed of 433 Mbps and a range of 20 m and Bluetooth 802.16 v4.0, with a transmission speed of 2 Mbps and a range of 10 m. For the message scheduling, a sender and a destination cast are selected uniformly and randomly from a set of nodes and predefined casts.

Table 5.1 Simulation parameters

Terrain dimension	4500 × 3400 m, segmented into 16 casts
Simulation time	57,600 s
Warm-up and cool-down periods for every simulation	2 h each
Buffer sizes	10 (default), 5, 15, 20, 25, 30, 35, 40 (Mb)
Message lifetimes	120 (default), 30, 60, 90, 150, 180, 210, 240 (min)
Message payload	500 KB
Message generation	Every 25 to 35 s
Scheduling policy	Random
Number of nodes	195 (default), 126, 189, 252, 315, 378, 441, 504, 567
Node speed	0.5–1.5 m/s

Other simulation parameters are given in Table 5.1.

5.4 Simulation Results

5.4.1 SPMBM Model

Figure 5.1 shows that as the number of hosts is increased, the delivery ratio increases for all the studied protocols. This is due to the fact that the more the number of nodes in the network, the better the chance that the message be delivered. In terms of delivery ratio performance, RLFGRP is about 4.4% better than MARL-CC and 16.6% better than FCSG.

Figure 5.2 reveals that for all the studied protocols, the average latency decreases when the number of hosts is increased. This is the direct consequence of the increase in delivery ratio. It is also observed that in terms of average latency, RLFGRP outperforms the other protocols. Indeed, in terms of average latency performance, RLFGRP scheme is about 2.42% better than MARL-CC and 7.6% better than FCSG.

Figure 5.3 shows that when the buffer size of nodes is increased, the delivery ratio increases. This is due to the fact that as the buffer size increases, more messages are stored in a node, leading to a better delivery ratio of messages to their destinations. Indeed, in terms of delivery ratio performance, RLFGRP is about 3.6% better than MARL-CC and 7.5% better than FCSG.

Figure 5.4 shows that as the buffer size is increased, the average latency increases. This is due to the fact the messages tend to stay longer than expected in the node's buffer, causing some delay in its delivery to the destination. It is also observed that in terms of average latency performance, RLFGRP is about 5.9% better than MARL-CC and 17.9% better than FCSG.

Figure 5.5 shows that when the TTL is increased, the delivery ratio also increases, and this increase is more pronounced for RLFGRP. This is due to the fact that the

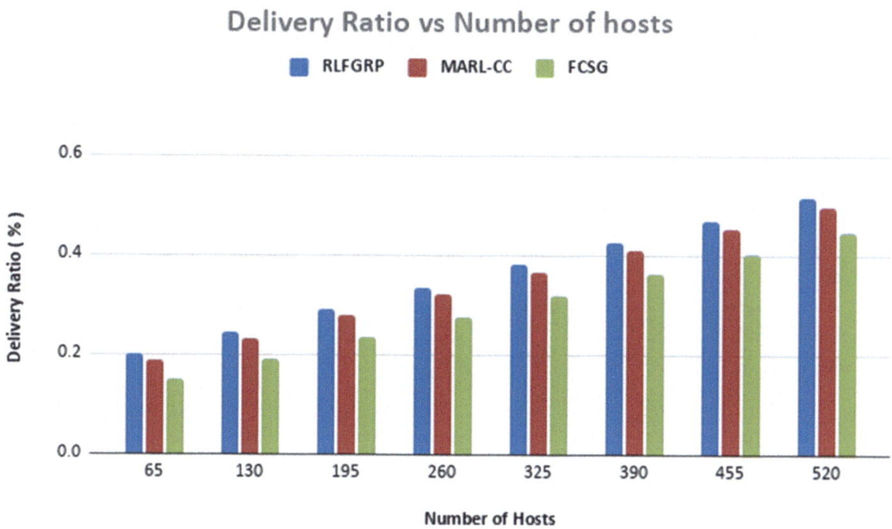

Fig. 5.1 Delivery ratio vs. number of hosts

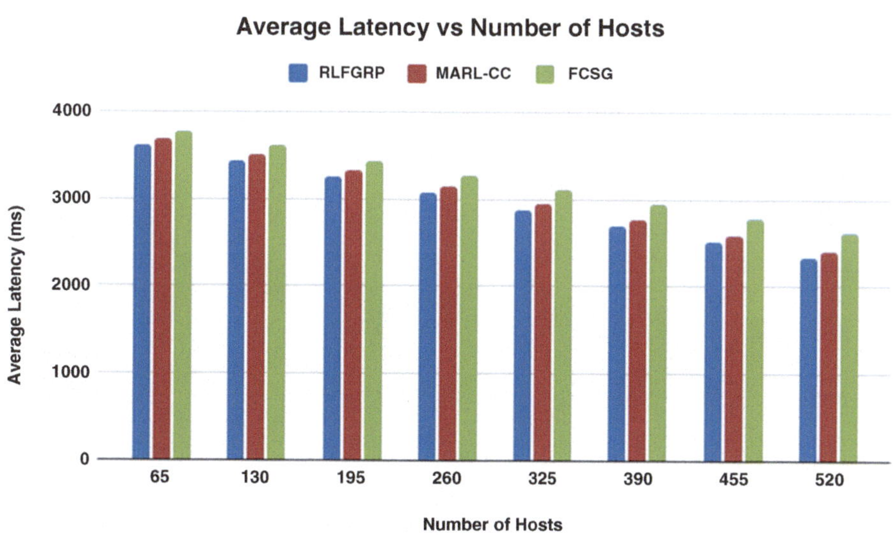

Fig. 5.2 Average latency ratio vs. number of hosts

messages get more time to find a suitable relay node to forward the message towards its destination. It is also observed that in terms of delivery ratio performance, RLFGRP scheme is about 2.4% better than MARL-CC and 13.6% better than FCSG.

In Fig. 5.6, it is observed that as the TTL increases, the average latency also increases for all the studied protocols, and this increase is less pronounced for

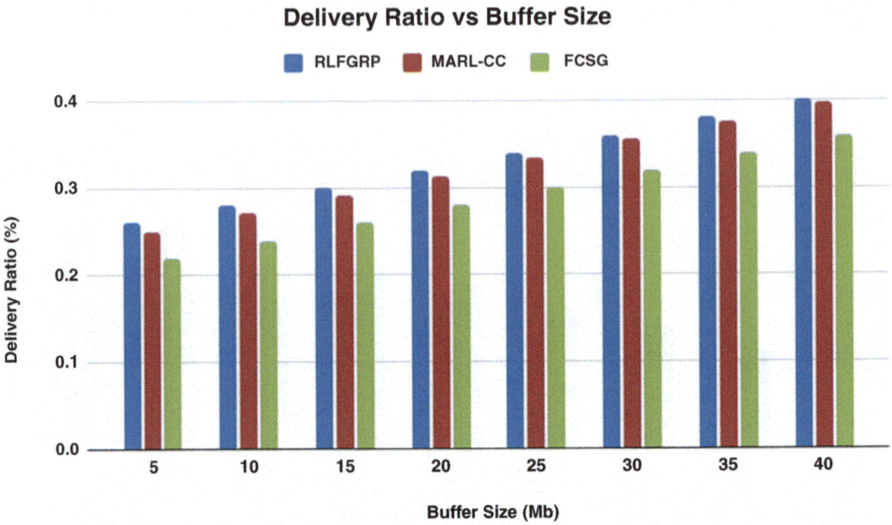

Fig. 5.3 Delivery ratio vs. buffer size

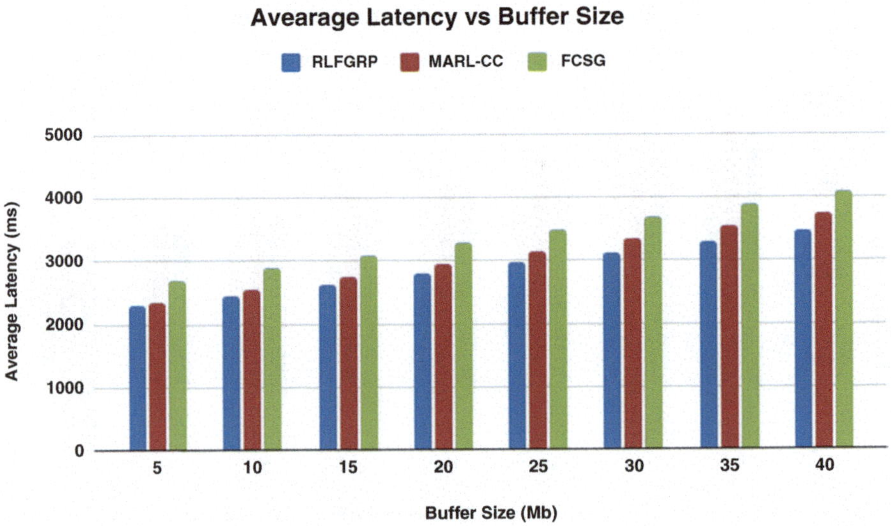

Fig. 5.4 Average Latency vs. buffer size

RLFGRP. This may be due to the fact that as the TTL of messages is increased, the number of forwarded messages also increases. In terms of average latency performance, it is found that RLFGRP is about 4.8% better than MARL-CC and 20.2% better than FCSG.

Figure 5.7 shows that for all studied protocols, as the number of hosts increases, the overhead ratio also increases. But this increase is less pronounced in the case

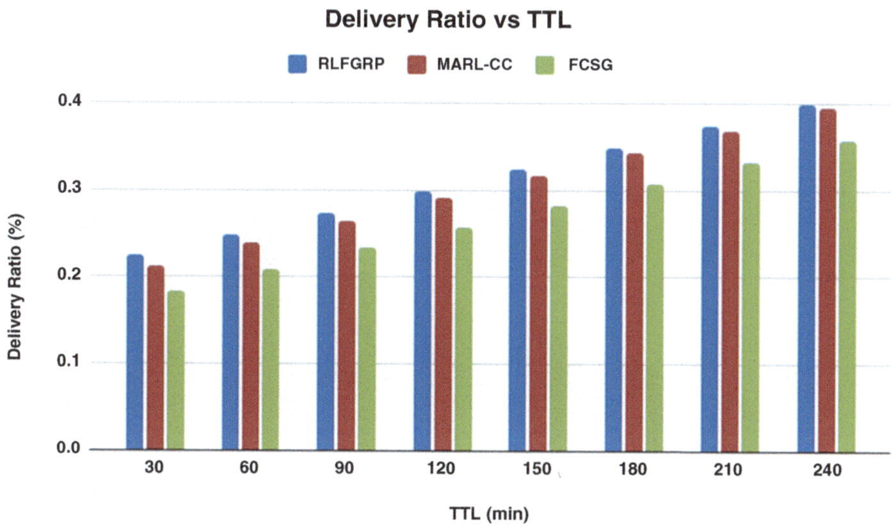

Fig. 5.5 Delivery Ratio vs. TTL

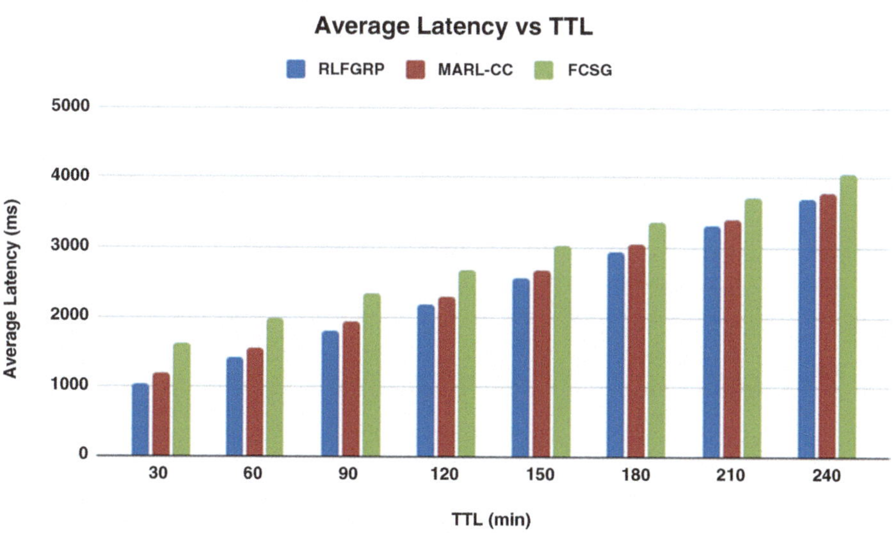

Fig. 5.6 Average latency vs. TTL

of RLFGRP. This might be attributed to the Check-and-Spray controlled flooding mechanism implemented in RLFGRP, which primarily imposes a threshold on the number of remaining message copies in the network to control the overhead reduction, by ensuring that the message once in the geocast region will not be spread outside that region. It is also observed that in terms of overhead ratio performance, RLFGRP is about 6.1% better than MARL-CC and 18.43% better than FCSG.

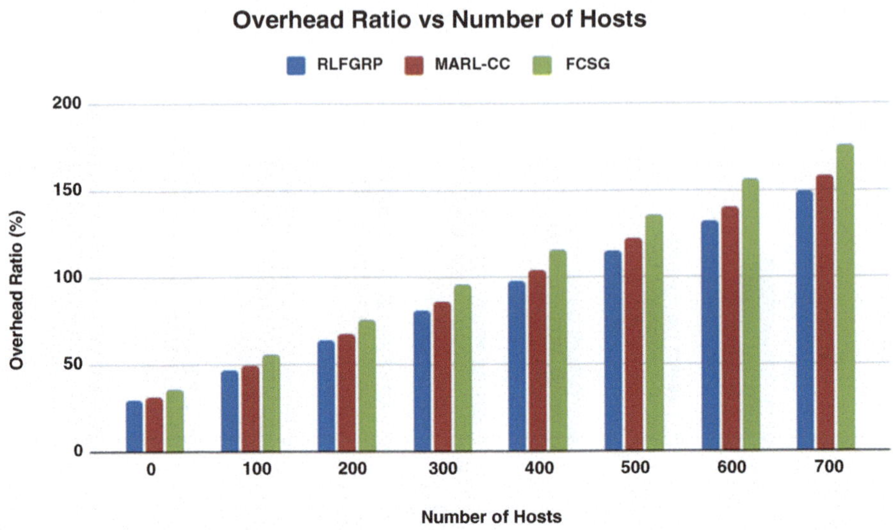

Fig. 5.7 Overhead ratio vs. number of hosts

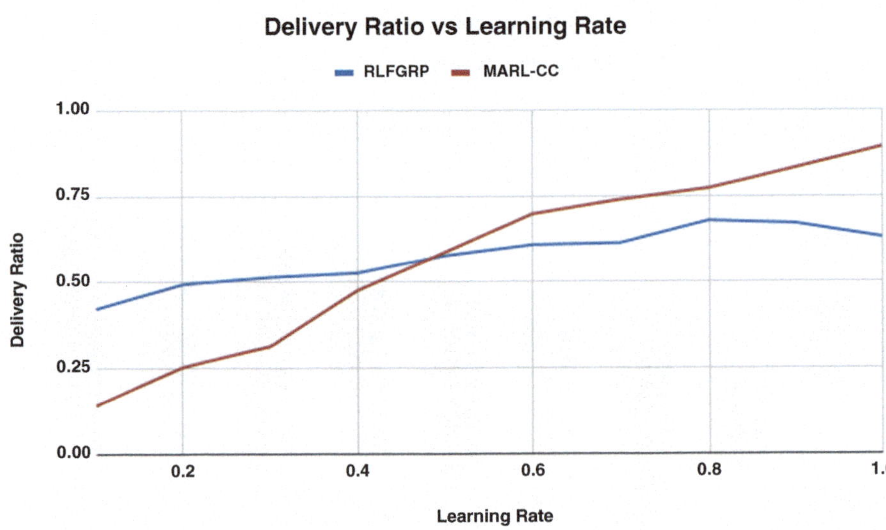

Fig. 5.8 Delivery ratio vs. learning rate

Figure 5.8 shows the effect of delivery ratio on the learning rate for the RLFGRP and MARL-CC learning algorithms. It is observed that RLFGRP performs well when the learning rate hits 0.8, whereas MARL-CC performs well when the learning rate hits 1.

5.4.2 Real Mobility Traces Model

In Fig. 5.9, it is found that all the studied protocols performed well when the buffer resources are sufficient and the delivery ratio increases as the buffer size is increased. It is also observed that in terms of delivery ratio performance, RLFGRP is about 1.4% better than MARL-CC and 6.85% better than FCSG.

In Fig. 5.10, it is observed that as the buffer size is increased, the average latency increases. It is also observed that in terms of average latency performance, RLFGRP is about 5.5% better than MARL-CC and 14.1% better than FCSG.

Figure 5.11 shows that when the TTL is increased, the delivery ratio also increases, and this increase is more pronounced for RLFGRP. This is due to the fact that the nodes get more time to find a suitable forwarder to carry the message towards its destination. It is also observed that in terms of delivery ratio performance, RLFGRP is about 2.0% better than MARL-CC and 15.7% better than FCSG.

In Fig. 5.12, it is observed that as the TTL increases, the average latency also increases for all studied protocols, and this increase is less pronounced for RLFGRP. It is also found that in terms of average latency performance, RLFGRP scheme is about 5.4% better than MARL-CC and 19.5% better than FCSG.

Figure 5.13 shows that for all the studied protocols, as the buffer size increases, the overhead ratio also increases. Moreover, in terms of overhead ratio performance, RLFGRP is about 5% better than MARL-CC and 11.5% better than FCSG.

Figure 5.14 shows that for all studied protocols, as the buffer size increases, the overhead ratio also increases. It is also observed that in terms of overhead ratio performance, RLFGRP is about 5.1% better than MARL-CC and 11.6% better than FCSG.

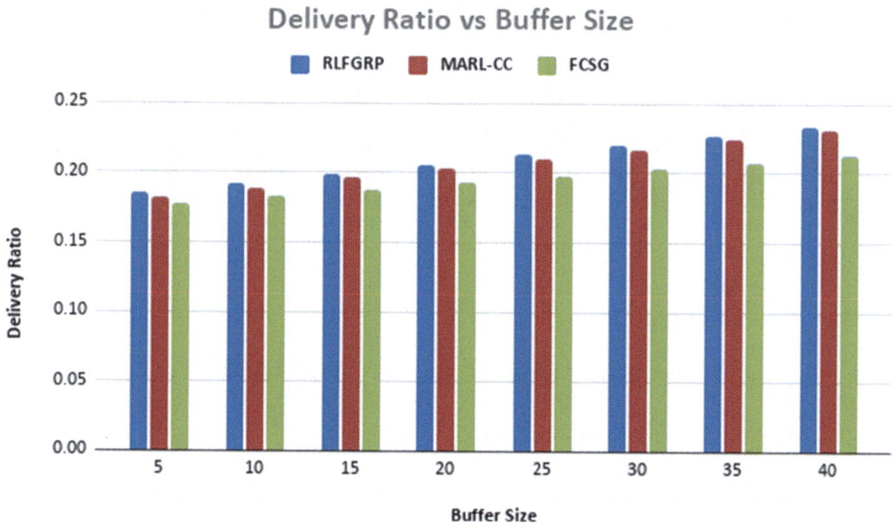

Fig. 5.9 Delivery ratio vs. buffer size

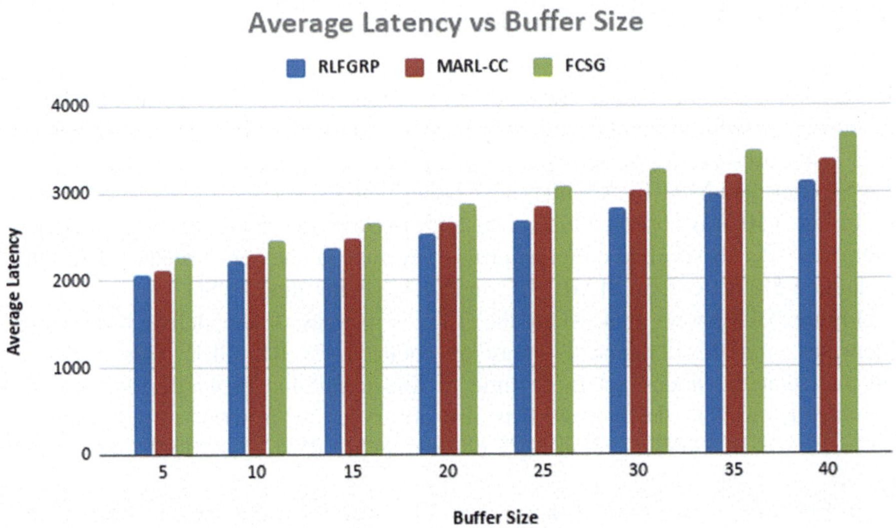

Fig. 5.10 Average latency vs. buffer size

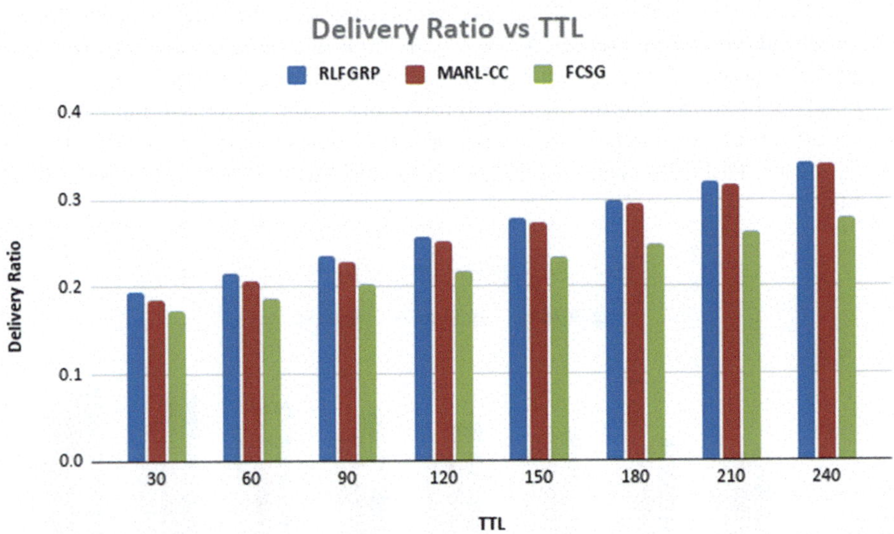

Fig. 5.11 Delivery ratio vs. TTL

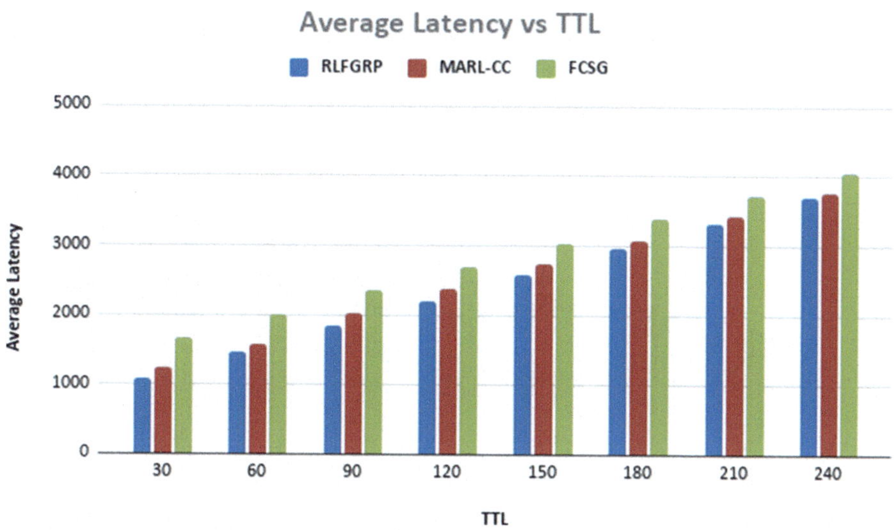

Fig. 5.12 Average latency vs. TTL

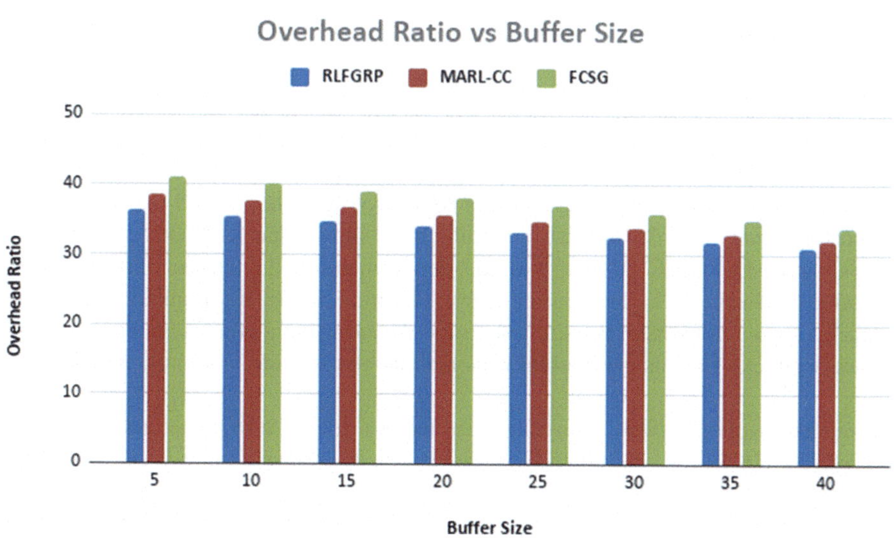

Fig. 5.13 Overhead ratio vs. buffer size

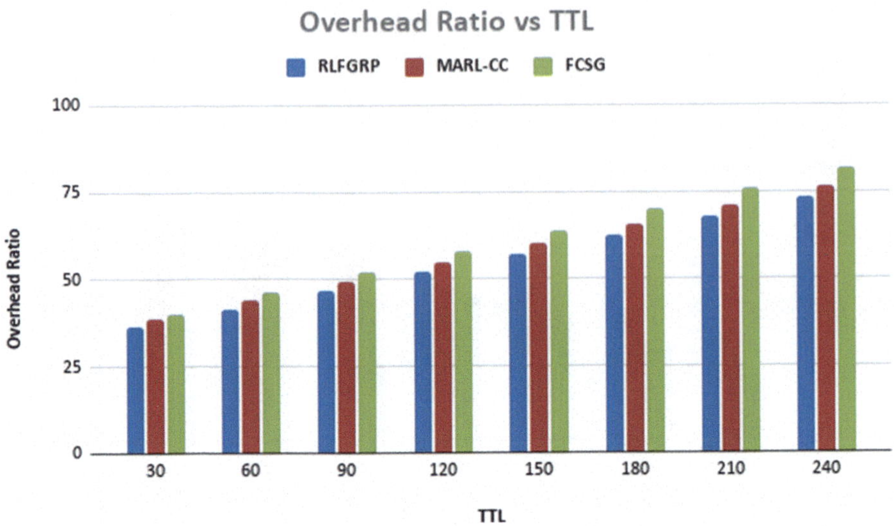

Fig. 5.14 Overhead ratio vs. TTL

5.5 Conclusion

In this chapter, we have compared three recently proposed fuzzy-based routing protocols for OppNets, namely RLFGRP, FCSG and MARL-CC, using the ONE simulator [15], under both the SPMBM model and the INFOCOM 2006 real mobility traces, considering the delivery ratio, average latency, and overhead ratio, as performance metrics. Simulation results have shown that RLFGRP outperforms FCSG and MARL-CC in terms of the above-mentioned metrics, under varying number of hosts, TTL, and buffer size. As future work, we plan to design the security-aware versions of the studied protocols and compare their resiliency against network attacks such as sybil attacks, wormhole attacks, and blackhole attacks.

Acknowledgments This work is partially sponsored by a grant held by the second author, from the National Science and Engineering Research Council of Canada (NSERC), reference number: RGPIN-2017-04423.

References

1. Lilien, L., Kamal, Z. H., Bhuse, V., & Gupta, A. (2007). The concept of opportunistic networks and their research challenges in privacy and security. In *Mobile and wireless network security and privacy* (pp. 85–117). Springer.
2. Huang, C. M., Lan, K. C., & Tsai, C. Z. (2008, March). A survey of opportunistic networks. In *22nd International Conference on Advanced Information Networking and Applications-Workshops (AINA Workshops 2008)* (pp. 1672–1677). IEEE.

3. Khalid, K., Woungang, I., Dhurandher, S. K., & Singh, J. (2020). *Reinforcement learning-based fuzzy geocast routing protocol for opportunistic networks*. Accepted Feb. 25, 2021.
4. Khalid, K., Woungang, I., Dhurandher, S. K., Singh, J., & Barolli, L. (2021). A fuzzy-based check-and-spray geocast routing protocol for opportunistic networks. *Journal of High Speed Networks, 27*(1), 1–12.
5. Singh, J., Dhurandher, S. K., & Woungang, I. (2020, December). Multi-agent reinforcement learning based efficient routing in opportunistic networks. In *2020 IEEE 17th India Council International Conference (INDICON)* (pp. 1–6). IEEE.
6. Kuppusamy, V., Thanthrige, U. M., Udugama, A., & Förster, A. (2019). Evaluating forwarding protocols in opportunistic networks: trends, advances, challenges and best practices. *Future Internet, 11*(5), 113.
7. Rajaei, A., Chalmers, D., Wakeman, I., & Parisis, G. (2016, June). GSAF: Efficient and flexible geocasting for opportunistic networks. In *2016 IEEE 17th International Symposium on a World of Wireless, Mobile and Multimedia Networks (WoWMoM)* (pp. 1–9). IEEE.
8. Dhurandher, S. K., Singh, J., Woungang, I., Takizawa, M., Gupta, G., & Kumar, R. (2019, November). Fuzzy geocasting in opportunistic networks. In *International Conference on Broadband and Wireless Computing, Communication and Applications* (pp. 279–292). Springer.
9. Rahimi, S., & Jamali, M. A. J. (2019). A hybrid geographic-DTN routing protocol based on fuzzy logic in vehicular ad hoc networks. *Peer-to-Peer Networking and Applications, 12*(1), 88–101.
10. Jain, S., Chawla, M., Soares, V. N., & Rodrigues, J. J. (2016). Enhanced fuzzy logic-based spray and wait routing protocol for delay tolerant networks. *International Journal of Communication Systems, 29*(12), 1820–1843.
11. Mottaghinia, Z., & Ghaffari, A. (2018). Fuzzy logic based distance and energy-aware routing protocol in delay-tolerant mobile sensor networks. *Wireless Personal Communications, 100*(3), 957–976.
12. Banerjee, A., & Dutta, P. (2008). Fuzzy controlled adaptive geocasting in mobile ad hoc networks. *International Journal of Information Technology, 14*(2), 109–131.
13. Khalid, K., Woungang, I., Dhurandher, S. K., Singh, J., & Rodrigues J. J. P. C. (2020). Energy-efficient check and spray geocast routing protocol for opportunistic networks. *Information, 11*, 504, 1–18. https://doi.org/10.3390/info11110504
14. Scott, J., Gass, R., Crowcroft, J., Hui, P., Diot, C., & Chaintreau, A. (2009). CRAWDAD dataset Cambridge/haggle (v. 2009-05-29). CRAWDAD Wireless Network Data Archive.
15. Keränen, A., Ott, J., & Kärkkäinen, T. (2009, March). The ONE simulator for DTN protocol evaluation. In *Proceedings of the 2nd International Conference on Simulation Tools and Techniques* (pp. 1–10).

Chapter 6
ACIDS: A Secure Smart City Framework and Threat Model

Soomaiya Hamid and Narmeen Zakaria Bawany

6.1 Introduction

More than 50% of today's world population dwells in urban areas to improve their quality of life, and this percentage is increasing with time [1]. The citizens' needs can be fulfilled efficiently if there is a well-established interconnected system to manage, maintain, and monitor the activities of the inhabitants [2]. Smart city is a human-friendly and efficient society that provides the core infrastructure for a quality life in almost all city-related facilities through smart city applications [3, 4]. These cities embrace information and communication technologies (ICT) to improve the quality and performance of civic services such as energy, municipal, health, transportation, safety, security, and utilities. Paroutis [5] affirmed that ICT involvement in urban services minimizes resource consumption, wastage, and overall cost. The smart city provisions an environment to connect all stakeholders and institutions by enabling intelligent and sustainable technologies and platforms like the Internet of Things (IoT) and cloud services. This promotes more efficient, convenient, and synchronized operations of urban infrastructure [6]. Therefore, smart cities have acquired more attention in the development and maintenance of a modern city [3].

Despite these benefits of interconnectivity and transparency, a smart city is prone to vulnerabilities and ultimately cybersecurity attacks. Smart city systems gather data from various sources, which include stakeholders and sensors, for the betterment of society. However, this sharing of data opens the opportunity for attackers to target a particular stakeholder or the entire system [7].

S. Hamid (✉) · N. Z. Bawany
Center for Computing Research, Department of Computer Science and Software Engineering, Jinnah University for Women, Karachi, Pakistan
e-mail: soomaiya.hamid@juw.edu.pk; narmeen.bawany@juw.edu.pk

I. Woungang, S. K. Dhurandher, *4th International Conference on Wireless, Intelligent and Distributed Environment for Communication*, Lecture Notes on Data Engineering and Communications Technologies 94, https://doi.org/10.1007/978-3-030-89776-5_6

Many cases have been reported in this regard as shown in Table 6.1. In early 2018, Atlanta was the victim of a virus named SamSam that severely affected their government agencies, hospitals, and big retailers. Colorado Department of Transportation of Atlanta (CODT) reported that SamSam shutdown more than 20,000 computers and pushed them into the dark ages where people had pen and paper to do their daily business [8]. Victims were then asked for bitcoins to get their files back. The SamSam was also used to attack Indiana [9] which disrupted the US hospital management system by encrypting the files and renamed them with the phrase "I'm sorry." The hospital management operations were halted for 2 days. This system was restored after paying 4 bitcoins worth $55,000 but took many days to smooth the hospital management operations.

In Czech Republic,[1] a cyberattack was triggered during the COVID-19 pandemic. Brno University Hospital's smart system was hacked which was one of the largest coronavirus centers in the Czech Republic. The attack paralyzed the whole IT system of the hospital. Attackers announced publicly that all surgeries are canceled. Hospital management failed to operate the COVID-19 testing system, and patients were shifted to other hospitals.

In 2018, Marriott[2] reported that data of 383 million travelers have been compromised in "a breach of Marriott's Starwood Preferred Guest (SPG) database." The investigation reveals that this data breach happened because of a few unencrypted passport numbers. Moreover, the report said that the attacker had unauthorized access since 2014. This attack not only revealed the payment details but revealed personal sensitive information. In 2015, Fiat Chrysler Automobiles declared that their Jeep Cherokee has been hacked by cybercriminals [10]. Therefore, the company had to recall 1.4 million cars. Afterward, the company had to install a security patch in every vehicle physically to secure the system.

In South Carolina [10], a mother noticed that the baby video monitor is moving around the room instead of focusing on the baby bassinet. First, she thought that some family member was controlling it with a smartphone app, but later she realized that it is being hacked and someone is collecting images of the personal activities.

The literature encompasses many smart city frameworks [11, 12], but most of the work is limited to an efficient interconnected architecture leaving behind its security aspect. However, there are specific smart city architectures that include security for a particular application only [13–15].

The rising number of attacks, along with the diversity in their types, clearly shows that we need new approaches and frameworks which prioritize the security aspect. Therefore, we proposed a layered smart city framework—ACIDS (Application, Communication, Infrastructure, Data, and Stakeholders)—that embeds security in

[1] https://www.zdnet.com/article/czech-hospital-hit-by-cyber-attack-while-in-the-midst-of-a-covid-19-outbreak/.

[2] https://news.marriott.com/2018/11/marriott-announces-starwood-guest-reservation-database-security-incident/ (accessed 10.14.2020).

Table 6.1 Popular attacks on smart city

Year	Target/victim	Affected ACIDS layer	Type of attack	Effect of attack
2020	Brno University Hospital management system, Czech Republic (https://www.zdnet.com/article/czech-hospital-hit-by-cyber-attack-while-in-the-midst-of-a-covid-19-outbreak/)	Communication layer	Ransomware attack	Hospital management failed to operate the COVID-19 testing system and patients' surgeries
2020	Worldwide Personal Privacy (https://www.who.int/about/communications/cyber-security)	Stakeholder layer	Phishing attack	The attacker sent fake invoices for requesting funds for COVID-19 victims, asked for personal information, posing himself as WHO (World Health Organization)
2019	New Orleans (https://www.pymnts.com/innovation/2020/secure-smart-cities-cybercriminals/)	Communication layer	Ransomware attack	Attack forced to the shutdown of thousands of systems of the city
2019	US hospital management system, Indiana (https://www.trendmicro.com/vinfo/tr/security/news/cyber-attacks/samsam-ransomware-hits-us-hospital-management-pays-55k-ransom)	Data layer	Ransomware attack	Attackers encrypted and renamed the files. The hospital management operations were halted for 2 days
2018	Government agencies, hospitals, and big retailers, Atlanta (https://www.iotworldtoday.com/2018/04/05/smart-city-security-atlanta-cyberattack-cripples-city/)	Infrastructure layer	Ransomware attack	Shutdown more than 20,000 computers
2018	Marriot Hotel, Maryland, USA (https://news.marriott.com/2018/11/marriott-announces-starwood-guest-reservation-database-security-incident/)	Data layer	Ransomware attack	Attack revealed the 383 million travelers' payment details and sensitive information

(continued)

Table 6.1 (continued)

Year	Target/victim	Affected ACIDS layer	Type of attack	Effect of attack
2016	USA (https://news.softpedia.com/news/a-massive-botnet-of-cctv-cameras-involved-in-ferocious-ddos-attacks-505722.shtml#ixzz4CsbxFc4A)	Application layer	DDoS attack	Attackers strive through CCTV-based botnet and occupied the server availability through illegitimate traffic by generating about 50,000 HTTP requests per second
2015	Fiat Chrysler Automobiles, North America (https://www.bbc.com/news/technology-33650491)	Communication layer	Remote hacking attack	A smart "Jeep Cherokee" has been hacked by cybercriminals
2015	Personal Security, South Carolina (https://www.npr.org/sections/thetwo-way/2018/06/05/617196788/s-c-mom-says-baby-monitor-was-hacked-experts-say-many-devices-are-vulnerable)	Infrastructure layer	Device hijacking	The baby video monitor was hacked, and the attacker was collecting images of the personal activities by moving the camera around the entire room instead of focusing on the baby bassinet
2015	Electric Power Grid, Ukraine (https://www.eenews.net/stories/1060040399/)	Infrastructure layer	Trojan horse	Attackers successfully hacked the power grid

each of its layers. Moreover, various threats respective to each layer and their consequences are presented in detail.

The major contributions of this chapter are as follows:

- We present a detailed comparison of existing smart city security frameworks.
- We propose a secure layered framework ACIDS for smart city.
- We present a threat model for a smart city that identifies major threats in each layer.

The rest of the chapter is organized as follows. Section 6.2 discusses the taxonomy of previous research work in this domain. ACIDS framework and its threat model are described in Sect. 6.3. Section 6.4 concludes the chapter and outlines future directions.

6.2 Related Literature

To address the cybersecurity threats in a smart city, an Anomaly Detection loT (AD-IoT) [16] system was proposed. AD-IoT intelligently detected anomalies by the Random Forest machine learning algorithm. This system also detected anomalies over compromised loT devices at distributed fog nodes. Researchers [17] have provided analysis and taxonomy of security and privacy challenges in the IoT layer only. Makhdoom et al. [18] present a blockchain-based framework "PrivySharing," which provides privacy and security of data sharing over a smart city network. Dong et al. [14] presented cyber issues in smart energy applications. Cyber-security challenges through a vulnerability assessment for the deployment of smart streetlight systems are presented. Brown and Seuwou [19, 20] presented smart city security and privacy challenges regarding mobility and transportation systems. Privacy and security challenges in smart healthcare are also discussed in various research [21–23]. Vitunskaite et al. [24] presented the role of IEEE standards and regulatory framework for cybersecurity with a comparative case study of three different countries. Many researchers [1, 25] presented various cybersecurity threats to smart city applications. To provide a secure platform for IoT devices, Chakrabarty and Engels [26] introduced four-block architecture.

Braun and Habibzadeh [21, 27] highlighted data privacy issues of a smart city and discussed the critical issues of cloud sharing platforms in a smart city. Furthermore, AlDairi and Tawalbeh [28] presented data privacy issues and smart city infrastructure challenges.

Cybersecurity challenges have been studied extensively in the smart city context [29]. We have examined more than 10 research papers in detail, and their findings are summarized in Table 6.2. Typically, previous studies are limited to a particular domain within a smart city, such as smart grid, smart traffic control system, VANETs, etc.

Table 6.2 A taxonomy of the research papers in security of smart city

Research paper	Smart city framework	Smart city framework with cybersecurity	Infrastructure layer security	Communication layer security	Data layer security challenges	Application layer security	Stakeholder layer security
[18]	✗	✗	✓	✗	✓	✗	✓
[19]	✗	✗	✗	✗	✗	✓	✗
[20]	✗	✗	✗	✗	✗	✓	✗
[16]	✗	✗	✓	✗	✗	✗	✗
[21]	✗	✗	✗	✗	✗	✓	✗
[25]	✗	✗	✓	✗	✗	✗	✗
[30]	✓	✗	✗	✗	✗	✗	✗
[27]	✗	✗	✗	✗	✓	✗	✗
[28]	✗	✗	✓	✗	✓	✗	✗
[31]	✗	✗	✗	✗	✗	✓	✗
[14]	✗	✗	✗	✓	✗	✓	✗
[26]	✗	✗	✗	✓	✓	✗	✗
[17]	✗	✗	✓	✓	✗	✗	✗
ACIDS	✓	✓	✓	✓	✓	✓	✓

In this study, we propose a layered model to cover all the domains of a smart city. Vulnerabilities corresponding to each layer are identified, and the threat model is also presented.

6.3 ACIDS: The Proposed Framework

Smart city applications are developed to improve the management of urban areas. A huge and complex network exists to control, maintain, and provide services in a smart environment. Thousands of sensors and IoT devices are deployed that generate a huge amount of data. The data is collected, processed, and analyzed by applications to provide various services to citizens. This creates a highly complex and tightly knitted architecture. To classify the relation among these attributes, this chapter introduced a layered framework for smart cities, titled ACIDS (Application, Communication, Infrastructure, Data, and Stakeholders) as shown in Fig. 6.1. The framework has five layers that represent the overall architecture of the smart city.

6.3.1 Infrastructure Layer

The infrastructure layer serves as a primary layer that constructs an entire framework to provide smart services to the citizens. This layer typically exists as a physical

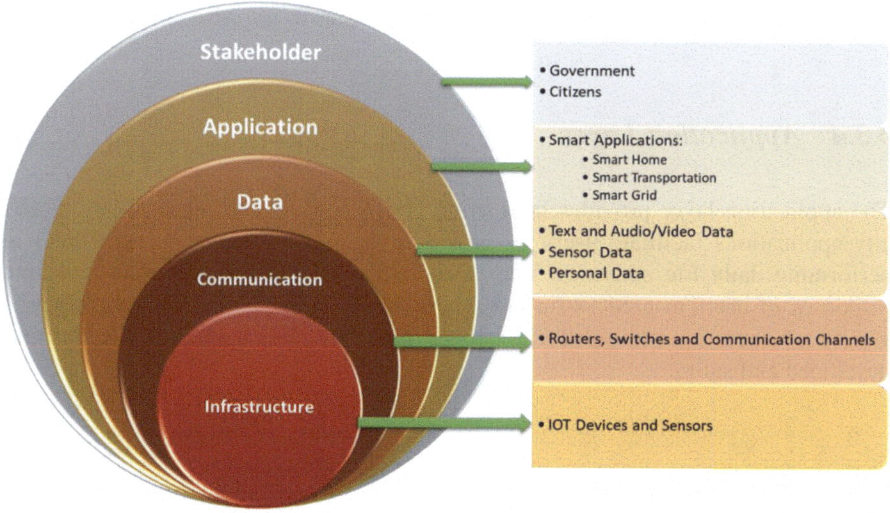

Fig. 6.1 Smart city layered model

platform that involves hardware such as actuators, IoT sensors, and other devices. This layer has a risk of physical damage or hijacking of the devices' control.

6.3.2 Communication Layer

This layer encompasses all the communication channels that can be used within a smart city. These communication channels include Wi-Fi, Ethernet, optical fiber, and broadband communications that are deployed across the smart city. Every device that is connected in smart city architecture needs strong communication to cover a wide geographical area. This huge and variable amount of data cannot communicate over a single communication technology. Therefore, multiple communication channels are used.

6.3.3 Data Layer

The IoT devices from the infrastructure layer generate a huge amount of data which includes structured and unstructured text, images, videos, and audios [32]. The data is generated from different smart city applications, such as transportation, utilities, health, business, energy, and waste management systems. Data layer carries big data platforms, to store, analyze, and process the data to provide ease to ICT projects of smart city. More data needs more computation power [33]. However, this data analysis plays an important role to build a city, smart. All applications of the smart city share data among them to provide better solutions to improve citizen's lives.

6.3.4 Application Layer

The application layer provides interaction between users and applications. Smart city applications facilitate users by providing ease and services to help them in performing daily life activities. This layer is responsible for collecting real-time responses of users to process further. These applications are developed for a vast variety of operations to solve city-related problems and help to make the city developed and safer.

6.3.5 Stakeholders

A stakeholder is a person or a group of persons that have a common interest in a system. They can either affect or be affected by the system. Smart cities help the government to provide a quality life to its citizens. Therefore, the two major categories of stakeholders that exist in a smart city are government and citizens. However, this part is least considered by the researchers of smart cities in their studies. It is important to emphasize that stakeholder roles must be established before developing any smart city plan because these players have the most influence on city initiatives and operations.

6.4 ACIDS Threat Model

A smart city provides complete connectivity among different sectors of modern society. Therefore, the data, services, and applications are integrated to build a strong smart city. This integrated nature of a smart city may attract many attackers to hack or disrupt the functions of a smart city, but due to its complex network, this becomes too difficult to identify which area of the network is vulnerable and prone to attacks. To overcome this difficulty, this research paper proposed an ACIDS threat model, which defines particular threats over each layer of ACIDS as shown in Fig. 6.2.

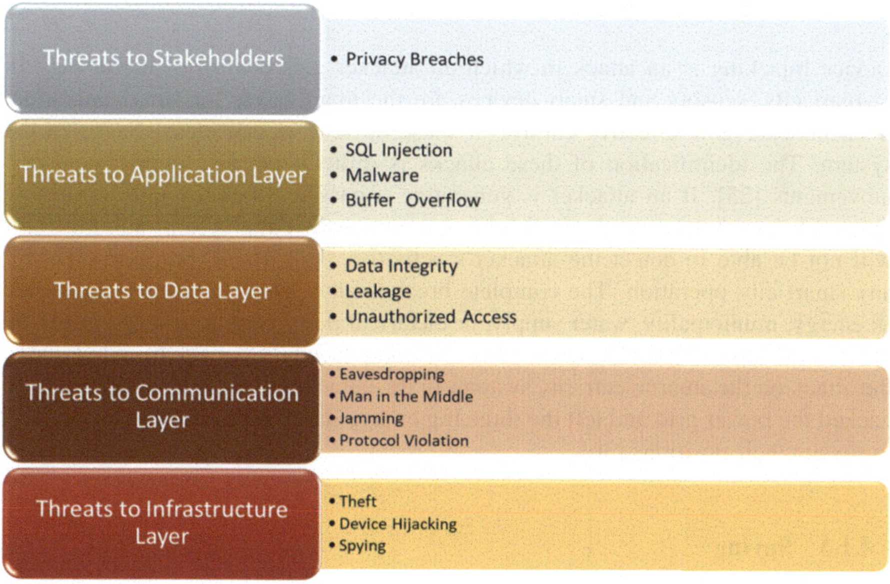

Fig. 6.2 ACIDS threat model

6.4.1 Threats to Infrastructure Layer

The infrastructure layer is a physical layer of smart city architecture. The layer can not only be compromised remotely but is also vulnerable to physical attacks. Therefore, the infrastructure layer has to be protected from both physical attacks and cyber-attacks. The following section discusses major attacks on the infrastructure layer, which defines how this layer can be targeted by attackers.

6.4.1.1 Theft

Theft is a very common attack that is performed by stealing tangible technological equipment. It affects the systems' availability and confidentiality. This kind of attack not only originates financial and reputational losses but also creates loopholes for attacks like impersonation and identity theft. In June 2019, [34] 20 laptops from the administration building of The University of Western Australia (UWA) were stolen. UWA reported that around 100,000 students, who have applied to study in the university from 1988 to 2018, are at risk due to this data breach.

6.4.1.2 Device Hijacking

Device hijacking is an attack in which an attacker gets control of the device. In a smart city, sensors and smart devices are the main assets for smart operation. If an attacker gets effective control of these devices, it can create havoc in the system. The identification of these attacks is difficult because of the attacker's movements [25]. If an attacker is generating a passive attack by only observing data and does not respond or alter basic functionality, the system administrator will not be able to detect the attacker's activities. This would be destructive for any smart city operation. The complete breakdown of smart city operations such as energy, municipality, water supply, or electrical power failure can be caused by device hijacking [35]. In December 2015, Ukraine faced a complete blackout due to the attack on the smart electricity system. In this attack, attackers have successfully hacked the power grid and left the three big energy distribution companies helpless to sustain their positions [36].

6.4.1.3 Spying

Security cameras are typically used in smart city applications for surveillance. A camera which is installed for security purposes becomes vulnerable when it is hacked and controlled by malicious users [37]. Attackers can get access to personal data and images or they can spy on people. If a camera is installed to cover the cashier's desk or at the banks where people use their cash and PINs fearlessly, a hacker could spy on people and plan a robbery [28]. Hackers can also replace the real-time streaming of a compromised camera with a tempered video or can completely block the video [38, 39].

6.4.2 *Threats to Communication Layer*

Communication layer keeps the smart city components interactive, by which devices and applications can communicate with each other. This layer is highly prone to attacks because it is exposed to all the layers of ACIDS. This layer is vulnerable to network traffic interception. These attacks may modify the communication to impersonate the user or service, or simply capture the communication channel so that they can perform malfunctions later with this information. Communication layer attackers also manipulate protocols to violate their rules and policies and create a way toward unauthorized access. A few of the communication layer attacks are described below:

6.4.2.1 Eavesdropping

Eavesdropping is an attack in which an attacker listens to all kinds of communication between users, applications, and communication channels. This unauthorized reader only reads the data without any interruption or tempering [40]. By eavesdropping, an attacker can perform a traffic analysis of confidential information about participants, pinpoint their location, or record their private conversations [41]. These kinds of attacks are not only threatening for smart applications but can also affect the privacy and security of all stakeholders.

6.4.2.2 Man in the Middle Attack

In cybersecurity Man-In-The-Middle (MITM) is a very common attack that takes an attacker one step forward from eavesdropping. Attack intercept communication among users and temper data during transmission. This may falsify the operators' actions and interrupt or spoof communication between two systems [42]. There are two phases to make the MITM attack successful; interception, and decryption.

Interception—in the first step, legitimate traffic is diverted to the attacker's network before reaching the destination. These attacks can be executed by creating free malicious Wi-Fi for the public. Once a victim connects with this unprotected network, the attacker gains full visibility of any online data exchange. Following are the few active approaches to intercept communication between two different nodes [43]. (a) IP Spoofing—is a technique in which an attacker alters the packet header to disguise himself as a legitimate application. As a result, when a user attempts to access that particular application, the attacker's website gets connected.

So that all the user's activities are shared with the attacker without consent. (b) ARP Spoofing—is an activity in which attackers disguise their own MAC address as a legitimate user's MAC address. The attacker generates a fake ARP message to inform the network that this MAC address is not linked with the user's IP address on a local area network. As a result, all data sent to that particular IP address is transmitted to the attacker's site. (c) DNS Spoofing—is a process of DNS cache poisoning, in which an attacker infiltrates a DNS server, redirecting the particular website address to its IP address. As a result, all users are directed to the fake site.

Decryption—once an attacker gets access to the user's communication data by interception, a two-way SSL communication traffic requires a process of decryption. Many methods exist for this purpose; few of them are discussed here. (a) HTTPS spoofing—is a technique in which a victim's browser receives a fake certificate after the interception phase. This certificate contains digital signatures associated with the compromised application. Therefore, the browser verifies the signatures from the existing list of trusted sites. As a result, the data is sent to the attacker's address from the victim's system. (b) SSL hijacking—is an activity that is performed during TCP handshake. The attacker shares forged authentication keys with the user and application both. This disguises a secure connection while the entire connection is under the control of the attacker. (c) SSL stripping—downgrades an HTTPS connection to HTTP by intercepting the TLS authentication sent from the application to the user. The attacker sends an unencrypted version of the application's site to the user while maintaining the secured session with the application. Meanwhile, the user's entire session is visible to the attacker.

6.4.2.3 Jamming

Jamming is one of the simplest attacks, which makes the communication channel occupied via malicious activities such that the legitimate nodes are unable to connect. The attacker generates interference signals to block communication channels and disrupt normal operations, due to which not only is performance degraded, but it also damages the control system. This attack mostly works effectively with wireless channels [44].

There are two categories of jamming attacks: active and reactive. Active Jammer's goal is to keep the channel busy regardless of whether the channel is being used or not. They continuously send strong radio signals which increase the noise interference at the receiver's side. Reactive Jammers notice the activity over the communication channel and send signals only when the channel is being used by legitimate users [45].

6.4.2.4 Protocol Violation

Ping of death attack—allows attacker attempts to crash, destabilize, or freeze the targeted smart system or service by sending malformed or oversized packets using a simple ping command. The Internet Protocol (IP) defines a maximum packet length of 65,536 bytes. Usually, networks do not support packets of that length. However, sending a ping packet larger than 65,535 bytes violates the Internet Protocol. Fragmentation occurs on larger packet sizes by splitting the packet into smaller chunks. When the target system attempts to reassemble the fragments and ends up with an oversized packet, a memory overflow could occur and lead to various system problems including the crash [46]. Ping of death attacks was particularly effective because the victim's identity could be easily spoofed. Also, an attacker would need no detailed knowledge of the machine he/she was attacking, except for its IP address.

Smurf Attack—is a type of Distributed Denial of Service attack (DDoS) in which a large number of Internet Control Message Protocol (ICMP) packets are broadcasted to the computer network. This malware generates a fake echo request containing a spoofed source IP, which is the target server IP address. As the request is broadcasted so every host connected to that network will respond with an echo ICMP packet to the spoofed server IP address. This amplifies the effect of the Smurf attack and makes the targeted server bring down. Due to this, network performance is degraded and servers become unavailable for legitimate traffic [29].

TCP SYN Flood Attack—exploits TCP three-way handshake to consume targeted server resources and render it unavailable for the entire network. TCP SYN attack behaves like a DDoS attack by sending TCP connection requests faster than the targeted machine can process [47].

6.4.3 *Threats to Data Layer*

A strong data sharing and dependency among smart city applications leads to issues of data security and privacy over smart cities. Attackers try to expose, destroy, alter, or steal data to generate further attacks. From unauthorized access, attackers target to disrupt the smart city operations.

6.4.3.1 Data and Identity Theft

Data is an important part of every ICT project or system, and the way it is stored and shared shows the security concerns of the administration. By default smart gadgets and devices generate unprotected data such as simple surveillance cameras, parking garages sensors, smart traffic controls, personal fitness gadgets, and so on. Due to the inheritance property of interconnectivity among smart city applications, an ample amount of data is shared by different applications [48]. This allows data to be used by other smart city applications, and attackers take advantage by making fraudulent transactions. Moreover, the attacker also learns from the previously shared data by the victim and uses it for impersonation [49].

6.4.3.2 Unauthorized Access Control

Most of the systems are initiated by capturing users' credentials. It gives the impression that this system is secure and no one can access it, except authorized users [50]. But unfortunately, this so-called secure system becomes vulnerable, when security protocols are not fully implemented leading to attacks such as weak authentication schema and tampering with authentication tokens.

6.4.3.3 Default or Test Accounts

Default accounts are often used to initiate a system for the first configuration. System administrators leave that account as it is and create more accounts to use the system. When an attacker discovers the installed software at the victim's side, it is quite easy to find out the default accounts of a standard system to login. If the system administrator did not remove the default account, the attacker gets access from this loophole [51]. Moreover, test accounts are created by developers during the development to test the system. Test accounts, if not deleted or disabled after deployment, create a backdoor for attackers [52].

6.4.4 Threats to Application Layer

The Application Layer of ACIDS plays an important role to build a bridge between users and computers. This is the first layer that can be affected by malware. Cybercriminals are constantly enhancing their abilities to approach new application layer threats. This layer includes an attack on applications and services smart services unavailable to legitimate users. Some common attacks are discussed below:

6.4.4.1 DoS and DDoS Attack

DoS attack is defined as a Denial of Service attack in which a server is flooded by illegitimate requests from a robotic client with TCP and UDP packets. Whereas a DDoS attack is a Distributed Denial of Service attack in which a server is targeted by multiple illegitimate clients from different regions. DDoS is more dangerous than DoS because of its distributed nature [53].

These multi-vector attacks boost the application layer to high risk by modifying their payload patterns continuously due to which the attack becomes more complex and undetectable. Application Flooding and Web server maximum threads are also types of Denial of Service attacks [54].

6.4.4.2 SQL Injection

SQL Injection (SQLi) is an attack that injects malicious SQL queries and executes them. SQL server controls the web application from the backend and does not want interference from outside the web application. To make sure of the security of the SQL server, developers apply security measures. But unfortunately, attackers bypass these security measures because of the vulnerabilities of the system [55]. The attacker injects SQL queries to show, add, modify, and delete records in the database.

After attacking the system by SQL injection, attackers can transfer data between application and database [56]. Due to this, the attacker pushes the device to compromise the security of the smart city by performing false operations [57].

6.4.4.3 Application Workflow

Many application developers have an assumption that the user will follow the application flow as designed. But attackers have a very different mindset to bypass these legal and smooth flow of an application. Many applications of smart cities are interconnected and transfer data to each other. Therefore, if an insecure application is fetching data from a secure one with a legal flow, a loophole is created through which an attacker can penetrate that secure application [58].

6.4.5 Threats to Stakeholders (TS)

The roles and responsibilities of stakeholders vary with their category, that is, government and citizens. Government plays an essential and critical role in a smart city to control and manage the city infrastructure and provide services to its citizens. Citizens' roles include all users of the smart city system.

Both citizens and government administrators are affected by any security breach or attack. The extent of damage/loss is dependent on the activity of an attacker, infected application, and the role of the stakeholder. If a smart home application is compromised, citizens will suffer more than the government. In contrast, if a smart taxation system is compromised and a hacker makes false entries, the government's revenue sheets are compromised.

Smart city services must incorporate cybersecurity solutions to identify and mitigate threats. This works best when cybersecurity becomes a part of the legal city plans. Singapore passed a bill to ensure that proactive steps must be followed by the operators to secure data and the infrastructure of a smart city [59]. The government of Singapore also initiated cybersecurity awareness programs in universities, government, and private sector institutions. Therefore, they are becoming a Smart Nation by developing a security mindset. Organizations in Singapore have to implement a cybersecurity regulatory framework that consists of policies and procedures to identify cybersecurity threats, and in case of any incident, they can report under law sections.

London's mayor has launched a "London City Challenge" to make London the world's best and unique smart city to live in [60]. With all other activities, he also invested in London Digital Security Centre [61]. This is a joint venture between the Mayor of London, the Metropolitan Police Service, and the City of London Police. By this effort, London protects its citizens and business from cyber-crimes on an enterprise level. They also created an Information Security cell to support their public bodies from credential thefts. Moreover, Hague Security Delta [62] is also serving more than 200 organizations in Europe by working together to establish a secure environment.

The involvement of all stakeholders either citizens or government is necessary to create a culture of cybersecurity across the smart city. The establishment of a crime-free ecosystem for a smart city can only be achieved by implementing security policies in the public and private sector organizations.

With all the possible anti-malware activities, sometimes only users become a backdoor for attracting attackers. Users can be tricked by attackers, by tempting them to click on malware to install via advertisements or popups. Novice users are more often trapped in fake and phishing certificate sites leading to security breaches and data leaks. Weak passwords are also one of the major sources to invite attackers. Typically, users set weak passwords as they are easy to recall, but dictionary and brute-force attacks can break them easily.

6.5 Conclusion

A smart city tends to improve the quality of life of its citizens by connecting all stakeholders, that is, government, community, and citizens. Although this connectivity is beneficial in various ways, it brings about many security challenges as it enhances the threat landscape. The strongly knitted smart city systems are more vulnerable to attacks. This research presents a layered framework for smart city security—ACIDS.

ACIDS is a layered architecture that segregates smart cities into five layers, that is, Infrastructure, Communication, Data, Application, and Stakeholders. This chapter also proposed an ACIDS threat model that identifies various threats and each layer, such that developers can incorporate an exclusive/specific security mechanism for each layer. The layered architecture proposed in this chapter is highly beneficial for developing secure smart city systems. The threat model presented in this chapter can help in reducing the vulnerabilities significantly.

This framework can be applied to various use cases of smart cities such as Smart Grid, Smart Water and Waste Management, Smart Transportation, etc. In the future, we would like to implement these systems using the proposed ACIDS framework along with the security mechanisms that protect from the threats at each layer.

References

1. Khatoun, R., & Zeadally, S. (2016). Smart cities: Concepts, architectures, research opportunities. *Communications of the ACM, 59*, 46–57. https://doi.org/10.1145/2858789
2. Hollands, R. G. (2008). Will the real smart city please stand up? Intelligent, progressive or entrepreneurial? *City, 12*, 303–320. https://doi.org/10.1080/13604810802479126
3. Su, K., Li, J., & Fu, H. (2011). Smart city and the applications. In *International conference on electronics, communications and control* (pp. 1028–1031).
4. Lazaroiu, G. C., & Roscia, M. (2012). Definition methodology for the smart cities model. *Energy, 47*, 326–332. https://doi.org/10.1016/j.energy.2012.09.028
5. Paroutis, S., Bennett, M., & Heracleous, L. (2014). A strategic view on smart city technology: The case of IBM smarter cities during a recession. *Technological Forecasting and Social Change, 89*, 262–272. https://doi.org/10.1016/j.techfore.2013.08.041
6. Nam, T., & Pardo, T. A. (2011). Conceptualizing smart city with dimensions of technology, people, and institutions. In *ACM international conference proceeding series* (pp. 282–291). ACM Press.
7. Elmaghraby, A. S., & Losavio, M. M. (2014). Cyber security challenges in smart cities: Safety, security and privacy. *Journal of Advanced Research, 5*, 491–497. https://doi.org/10.1016/j.jare.2014.02.006
8. Kraszewski, K. (2019). SamSam and the silent battle of Atlanta. In *International conference on cyber conflict, CYCON*. NATO CCD COE Publications.
9. Zimba, A., & Chishimba, M. (2019). On the economic impact of crypto-ransomware attacks: The state of the art on enterprise systems. *European Journal for Security Research, 4*, 3–31. https://doi.org/10.1007/s41125-019-00039-8
10. Williams, M. (2015). 1.4 million cars that could be hacked are recalled by Fiat Chrysler In *PCWorld from IDG*. https://www.pcworld.com/article/2952592/chrysler-recalls-14m-cars-that-were-vulnerable-to-remote-hacking.html
11. Jin, J., Gubbi, J., Marusic, S., & Palaniswami, M. (2014). An information framework for creating a smart city through internet of things. *IEEE Internet of Things Journal, 1*, 112–121. https://doi.org/10.1109/JIOT.2013.2296516
12. Bawany, N., & Shamsi, J. (2015). Smart city architecture: Vision and challenges. *International Journal of Advanced Computer Science and Applications, 6*(11), 246–255. https://doi.org/10.14569/ijacsa.2015.061132
13. Ding, D., Conti, M., & Solanas, A. (2016). A smart health application and its related privacy issues. In *Smart city security and privacy workshop* (pp. 11–15). IEEE.
14. Jin, D., Hannon, C., Li, Z., et al. (2016). Smart street lighting system: A platform for innovative smart city applications and a new frontier for cyber-security. *The Electricity Journal, 29*, 28–35. https://doi.org/10.1016/j.tej.2016.11.011
15. Khurana, H., Hadley, M., Lu, N., & Frincke, D. A. (2010). Smart-grid security issues. *IEEE Security and Privacy, 8*, 81–85. https://doi.org/10.1109/MSP.2010.49
16. Alrashdi, I., Alqazzaz, A., Aloufi, E., et al. (2019). AD-IoT: Anomaly detection of IoT cyberattacks in smart city using machine learning. In *9th annual computing and communication workshop and conference* (pp. 305–310). IEEE.
17. Babar, S., Mahalle, P., Stango, A., et al. (2010). Proposed security model and threat taxonomy for the Internet of Things (IoT). In *Communications in computer and information science* (pp. 420–429). Springer.
18. Makhdoom, I., Zhou, I., Abolhasan, M., et al. (2020). PrivySharing: A blockchain-based framework for privacy-preserving and secure data sharing in smart cities. *Computers & Security, 88*, 101653. https://doi.org/10.1016/j.cose.2019.101653
19. Brown, M. (2020). Smart transport. In *Smart cities in application* (pp. 802–813). Springer.
20. Seuwou, P., Banissi, E., & Ubakanma, G. (2020). *The future of mobility with connected and autonomous vehicles in smart cities* (pp. 37–52). Springer.

21. Habibzadeh, H., & Soyata, T. (2019). Toward uniform smart healthcare ecosystems: A survey on prospects, security, and privacy considerations. In *Connected health in smart cities* (pp. 75–112). Springer.
22. Ranjith, J., & Mahantesh, K. (2019). Privacy and security issues in smart health care. In *4th international conference on electrical, electronics, communication, computer technologies and optimization techniques, ICEECCOT 2019* (pp. 378–383). Institute of Electrical and Electronics Engineers Inc.
23. Zeadally, S., Siddiqui, F., Baig, Z., & Ibrahim, A. (2019). Smart healthcare: Challenges and potential solutions using internet of things (IoT) and big data analytics. *PSU Research Review, 4*, 149–168. https://doi.org/10.1108/prr-08-2019-0027
24. Vitunskaite, M., He, Y., Brandstetter, T., & Janicke, H. (2019). Smart cities and cyber security: Are we there yet?A comparative study on the role of standards, third party risk management and security ownership. *Computers & Security, 83*, 313–331. https://doi.org/10.1016/j.cose.2019.02.009
25. Chen, J., Zuo, C., Diao, W., et al. (2019). Your IoTs are (not) mine: On the remote binding between IoT devices and users. In *IEEE/IFIP international conference on dependable systems and networks* (pp. 222–233). IEEE.
26. Chakrabarty, S., & Engels, D. W. (2016). A secure IoT architecture for Smart Cities. In *Annual consumer communications and networking conference* (pp. 812–813). IEEE.
27. Qamar, T., Bawany, N. Z., Javed, S., & Amber, S. (2019). Smart city services ontology (SCSO): Semantic modeling of smart city applications. In *Proceedings - 2019 7th international conference on digital information processing and communications, ICDIPC 2019* (pp. 52–56). Institute of Electrical and Electronics Engineers Inc.
28. Braun, T., Fung, B. C. M., Iqbal, F., & Shah, B. (2018). Security and privacy challenges in smart cities. *Sustainable Cities and Society, 39*, 499–507. https://doi.org/10.1016/j.scs.2018.02.039
29. Aldairi, A., & Tawalbeh, L. (2017). Cyber security attacks on smart cities and associated mobile technologies. *Procedia Computer Science, 109*, 1086–1091. https://doi.org/10.1016/j.procs.2017.05.391
30. Khatoun, R., & Zeadally, S. (2017). Cybersecurity and privacy solutions in smart cities. *IEEE Communications Magazine, 55*, 51–59.
31. Bawany, N., & Shamsi, J. (2019). SEAL: SDN based secure and agile framework for protecting smart city applications from DDoS attacks. *Journal of Network and Computer Applications, 145*, 102381. https://doi.org/10.1016/J.JNCA.2019.06.001
32. Chiesa, G. (2020). *Data, properties, smart city* (pp. 169–190). Springer.
33. Chamoso, P., González-Briones, A., De La Prieta, F., et al. (2020). Smart city as a distributed platform: Toward a system for citizen-oriented management. *Computer Communications, 152*, 323–332. https://doi.org/10.1016/j.comcom.2020.01.059
34. Hiatt, B. (2019). UWA student data may be compromised after laptop theft—The West Australian. In *WA News*—Educ. https://thewest.com.au/news/wa/uwa-student-data-may-be-compromised-after-laptop-theft-ng-b881272938z
35. Al-Taleb, N., Saqib, N. A., Atta-ur-Rahman, & Dash, S. (2020). Cyber threat intelligence for secure smart city. arXiv.
36. Adepu, S., Kandasamy, N. K., Zhou, J., & Mathur, A. (2020). Attacks on smart grid: Power supply interruption and malicious power generation. *International Journal of Information Security, 19*, 189–211. https://doi.org/10.1007/s10207-019-00452-z
37. Mohammed, F., Idries, A., Mohamed, N., et al. (2014). UAVs for smart cities: Opportunities and challenges. In *International conference on unmanned aircraft systems, ICUAS* (pp. 267–273). IEEE Computer Society.
38. Valente, J., & Cárdenas, A. A. (2015). Using visual challenges to verify the integrity of security cameras. In *ACM international conference proceeding series* (pp. 141–150). Association for Computing Machinery.
39. Takefuji, Y. (2018). Connected vehicle security vulnerabilities [commentary]. *IEEE Technology and Society Magazine, 37*, 15–18.

40. Baig, Z. A., Szewczyk, P., Valli, C., et al. (2017). Future challenges for smart cities: Cyber-security and digital forensics. *Digital Investigation, 22*, 3–13. https://doi.org/10.1016/j.diin.2017.06.015
41. Qu, Y., Nosouhi, M. R., Cui, L., & Yu, S. (2018). Privacy preservation in smart cities. In *Smart cities cybersecurity and privacy* (pp. 75–88). Elsevier.
42. Dua, A., Kumar, N., Das, A. K., & Susilo, W. (2018). Secure message communication protocol among vehicles in smart city. *IEEE Transactions on Vehicular Technology, 67*, 4359–4373.
43. Maruˇ, M. (2016). Automatization of MitM attack for SSL/TLS decryption original connection. In *Excel@FIT 2016, student conference of innovation, technology and science in it* (pp. 2–8).
44. Niu, J., Ming, Z., Qiu, M., et al. (2015). Defending jamming attack in wide-area monitoring system for smart grid. *Telecommunication Systems, 60*, 159–167. https://doi.org/10.1007/s11235-014-9930-3
45. Garcia-Font, V., Garrigues, C., & Rifà-Pous, H. (2016). A comparative study of anomaly detection techniques for smart city wireless sensor networks. *Sensors, 16*, 868. https://doi.org/10.3390/s16060868
46. Abdollahi, A., & Fathi, M. (2020). An intrusion detection system on ping of death attacks in IoT networks. *Wireless Personal Communications*, 1–14. https://doi.org/10.1007/S11277-020-07139-Y
47. Kepceoglu, B., Murzaeva, A., & Demirci, S. (2019). Performing energy consuming attacks on IoT devices. In *27th telecommunications forum, TELFOR 2019*. Institute of Electrical and Electronics Engineers Inc.
48. Choenni, S., Bargh, M. S., Roepan, C., & Meijer, R. F. (2016). Privacy and security in smart data collection by citizens. In *Public administration and information technology* (pp. 349–366). Springer.
49. Ghosh, D., Ae Chun, S., Adam, N. R., & Shafiq, B. (2016). Big data-based smart city platform: Real-Time crime analysis. In *ACM international conference proceeding series* (pp. 58–66). ACM.
50. Dagher, G. G., Mohler, J., Milojkovic, M., & Marella, P. B. (2018). Ancile: Privacy-preserving framework for access control and interoperability of electronic health records using blockchain technology. *Sustainable Cities and Society, 39*, 283–297. https://doi.org/10.1016/j.scs.2018.02.014
51. Jain, A. K., & Shanbhag, D. (2012). Addressing security and privacy risks in mobile applications. *IT Professional, 14*, 28–33. https://doi.org/10.1109/MITP.2012.72
52. Yu, W. D., Supthaweesuk, P., & Aravind, D. (2005). Trustworthy Web Services based on testing. In *Proceedings - SOSE 2005: IEEE international workshop on service-oriented system engineering* (pp. 167–177).
53. Bawany, N., & Shamsi, J. (2016). Application layer DDoS attack defense framework for smart city using SDN. In *Proceedings of the third international conference on computer science, computer engineering, and social media* (pp. 1–9). IEEE.
54. Sonar, K., & Upadhyay, H. (2016). An approach to secure Internet of Things against DDoS. In *Proceedings of international conference on ICT for sustainable development* (pp. 367–376). Springer.
55. Kim, J. M., Jeong, H. Y., Cho, I., et al. (2014). A secure smart-work service model based OpenStack for cloud computing. *Cluster Computing, 17*, 691–702. https://doi.org/10.1007/s10586-013-0251-1
56. Ahamed, J., & Rajan, A. V. (2017). Internet of Things (IoT): Application systems and security vulnerabilities. In *International conference on electronic devices, systems, and applications*. IEEE Computer Society.
57. Choi, J., Spaulding, J., Anwar, A., et al. (2019). IoT malware ecosystem in the wild: A glimpse into analysis and exposures. In *Proceedings of the 4th ACM/IEEE symposium on edge computing, SEC 2019* (pp. 413–418). ACM.

58. Dewi Rosadi, S., Suhardi, S., & Kristyan, S. A. (2017). Privacy challenges in the application of smart city in Indonesia. In *2017 international conference on information technology systems and innovation, ICITSI 2017 - proceedings* (pp. 405–409). Institute of Electrical and Electronics Engineers Inc.
59. Hoe, S. L. (2016). Defining a smart nation: The case of Singapore. *Journal of Information, Communication and Ethics in Society, 14*, 323–333. https://doi.org/10.1108/JICES-02-2016-0005
60. Gupta, A., Panagiotopoulos, P., & Bowen, F. (2020). An orchestration approach to smart city data ecosystems. *Technological Forecasting and Social Change, 153*, 119929. https://doi.org/10.1016/j.techfore.2020.119929
61. Angelidou, M. (2017). The role of smart city characteristics in the plans of fifteen cities. *Journal of Urban Technology, 24*, 3–28. https://doi.org/10.1080/10630732.2017.1348880
62. Rothkrantz, L. J. M. (2016). Flood control of the smart city Prague. In *2016 smart cities symposium Prague, SCSP 2016*. Institute of Electrical and Electronics Engineers Inc.

Chapter 7
Volunteer Drone: Search and Rescue of the Industrial Building Collapsed Worker

A. K. M. Islam, Dalia Hanna, and Alexander Ferworn

7.1 Introduction

Efforts to produce machine-assisted disaster management practices are being implemented all across the world. Machines have become mobile enough, so as to replace humans in dangerous situations. Therefore, using robotic assistants is a new trend in conceiving and managing natural disaster management protocols. For example, in the case of the La Conchita mudslide in 2005, a shoebox-sized wheeled robot was inserted into the damaged house to scan for victims [1]. Furthermore, at the earthquake-hit Mirandola in 2012, a team of humans and robots (unmanned ground vehicle, unmanned aerial vehicle) worked together with the Italian National Fire Corps [1]. Even more recently, a field experiment was also conducted with a team of ground and aerial robots toward the mapping of an earthquake damaged building at 2011 Tohoku earthquake [1]. Among various types of robots, unmanned aerial vehicles, also known as drones, have been extensively developed to monitor and access disaster sites. Drones are cost-effective, compact, and easy to operate in a cluttered environment [2], and have thus become affordable candidates for disaster assistance.

It is extremely challenging to explore damaged industrial buildings due to low-light conditions and structural imbalance. In preparation for these conditions, drones can be fused with sensors to obtain 3D mapping of the unknown area so that the rescuers do not have to risk their lives exploring the location. Gathering information about the localization of victims and reaching them promptly is of the essence. In the real-world, however, it is not as easy to detect victims inside of a collapsed building. Two main problems are the lack of GPS coverage and lack of lighting. Sometimes

A. K. M. Islam (✉) · D. Hanna · A. Ferworn
Ryerson University, Toronto, ON, Canada
e-mail: akmzahidul.islam@ryerson.ca; dhanna@ryerson.ca; aferworn@ryerson.ca

I. Woungang, S. K. Dhurandher, *4th International Conference on Wireless, Intelligent and Distributed Environment for Communication*, Lecture Notes on Data Engineering and Communications Technologies 94, https://doi.org/10.1007/978-3-030-89776-5_7

objects obstruct the drone's field of vision and hinder the ability to detect human presence inside thick cement walls [2].

Recent studies on the use of drones in SAR missions utilize mobile phone detection technologies to locate individuals. Among other wireless technologies used for detection of mobile phones, there are 2/3/4G cellular networks, Wi-Fi, and Bluetooth [3]. Mobile phone detection technologies can be used jointly with a computer vision system, which improves their efficiency when there is a GPS denied environment or a presence of thick walls of cement or the limitation of light between the detector sensor and the lost person. This study has proposed a solution method of search and rescue inside of the collapsed industrial building by adding a BLE (Bluetooth Low Energy) detector with the infrared sensor-based camera with Lidar on drone about which is to be discussed in the rest of this chapter.

This chapter is organized as follows. In Sect. 7.2, the literature review is provided. In Sect. 7.3, proposed architecture of the drone assist disaster management system is provided. In Sect. 7.4, methodology is provided. In Sect. 7.5, result and analysis are provided. Finally, Section 7.6 concludes the chapter.

7.2 Literature Review

In recent years, autonomous drone platforms have received increased attention from the research community. Many authors have developed drone architectures capable of simultaneous localization and mapping (SLAM) [4–8], which became the basis for autonomous navigation in indoor flying scenarios. In particular, these architectures benefit UAV-assisted natural disaster management. While some researchers [8, 9] use GPS information for drone navigation, they propose to develop a drone architecture that is able to perform SLAM with onboard sensors. In general, there is no available GPS signal at disaster sites, and establishing sensor framework for navigating these unknown areas is crucial. Chen et al. [10] generated collision-free trajectories for drone flight using 3D grids in unknown environments. Camera-based [6, 7, 11] approaches known as visual SLAM also have been extensively studied to aid autonomous navigation. In [11], the authors fused laser readings with visual SLAM to robustly track aerial vehicle positions. Saska et al. in [12] developed groups of aerial vehicles localizing themselves in GPS-denied environments using visual relative localization. Although vision-based techniques can help provide pose estimation and 3D reconstruction, these approaches are difficult to implement in low-light conditions. Few works [4, 5] present autonomous indoor flying in GPS-denied areas, with a help of laser rangefinder sensor, which can resolve the lighting problems. Seoungjun Lee et al. [1] proposed to fuse IR depth camera with Lidar in order to provide local (nearby victims) and global (overall floor plan) maps of disaster area. The main limitation of this work is if there is a thick cement wall between the victims and the Lidar camera, this method is not reliable.

To overcome those limitations, the architecture extends this system and proposes a method of coordinate detection for living victims. One of the popular candidates in wireless technology for indoor positioning is Bluetooth Low Energy (BLE) [13]. The victim is considered a BLE-enabled device, and the study proposes a method to detect the distance of the devices from the drone. At the end of, this study will suggest a rescue method of the victim analysis by aggregating the data from the overall experiment. In this research, each UAV can be considered as a heterogeneous mobile gateway. BLE detection has been a successful project for different domain, but in this study, we will utilize this architecture for search and rescue operation using drones. This architecture allows for fast tracking of survivors, and thus, it will be effective in practical rescue operations.

7.3 Proposed Architecture of the Drone Assist Disaster Management System

This section introduces the architecture of the drone assist disaster management system. Most of the industrial buildings has a common structure according to the international industrial building construction code. So this proposed architecture will be enough for any industrial collapsed building. At the beginning of the experiment establish a control center besides the collapsed building to control the drone operation. Attach a BLE scanner and an infrared sensor-based camera with Light Detection and Ranging (Lidar) to the drone. First, the flying drone scans the collapsed building from the top using the BLE scanner and tries to find out BLE locations (latitude and longitude) inside of the building and produce a 2D map of the points. This will be used as the global map of the victim's position.

Among those points, some of the points are reachable by the drone and some are not. First, insert the drone inside of the collapsed building and try to search those reachable points and using the proposed mathematical equation (explain in Sect. 7.4.3.1) get the z value and build a 3D map of the victim's location. At the same time capture the drone's Lidar provides the 2D map of the surrounding points and utilize drone's camera to visualize local 3D map of the points and also add them to the 3D map.

Build a Mamdani Fuzzy Inference System (FIS) from those 3D values and implement that system to the non-reachable points and generate a 3D map for those points. Then add two 3D maps together and build a final 3D map of the victim's position inside of the collapsed building for further analysis of the rescue operation. This is a local map of the victim's position. Finally, cluster those 2D points of the 3D map using the K-means clustering algorithm and start rescue operation using the centroid of the clusters to rescue the survivors earliest possible time.

In the end, mining, which is finding a pattern for the similar scenario, of the drone trajectories provides the final outcome of the research. To do so, recursively implement a similar process to different scenarios, and a combination of a different result is the final result of the research. To evaluate this architecture, we implement the drone's behavior in an unknown scenario and predict the result. Finally, by

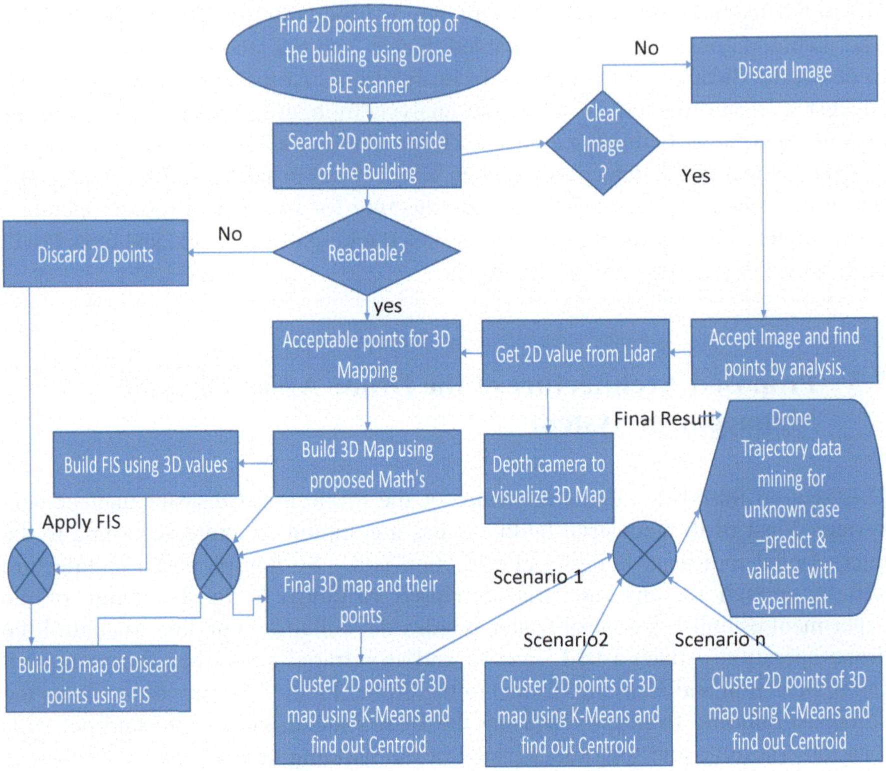

Fig. 7.1 Flowchart of the proposed drone assist disaster management system

comparing the predicted result to the practical experiment of that unknown scenario, we calculate the accuracy of the proposed system. The flow diagram in Fig. 7.1 illustrates the proposed drone assist disaster management system.

7.4 Methodology

The proposed architecture aims to detect survivors of industrial collapsed building, and speedy localization of the victims is the essential motivation of the overall framework. The overall project is divided into five segments:

1. Project configuration
2. 2D map generation of victim's position
3. Get 3D map using 2D map of victim's position
4. Cluster 3D map points using K-means clustering algorithms
5. Drone trajectory data mining

To sum up, the overall setup is able to construct global and local maps via the drone flight at the disaster environments and find the victims of natural disasters from an unknown building structures with limited lighting conditions or GPS-denied environment or a presence of thick walls of cement. They are described below step by step.

7.4.1 Project Configuration

7.4.1.1 Industrial Building Structure

Consider a scenario inside of the building where the total area needed to scan is about 2000 square feet and the height of the building is about 20 m (according to the International industrial building construction code). Initial efforts will use one drone for the scan.

7.4.1.2 Drone Structure

The study considers different DJI Mavic drones available in the market and compares their functionality, weight, automatic obstacle avoidance, accessibility, and life time (Table 7.1).

The Mavic Air 2 drone with propellers will be used to conduct the experiment with internal BLE scanner, built-in IMU, Lidar, and IR depth camera attached to it.

7.4.1.3 Control Center

The rescue control center collects necessary information regarding the disaster sites and oversees the entire rescue procedure. Any image or BLE data obtained by drone throughout the rescue mission can be visualized on the screen of the control center for analysis. Steps are described below:

- Camera-based target detection and positioning system.

Table 7.1 Comparison of different Mavic drones

Drone	Camera image quality	Automatic obstacle avoidance	Can get closer to area of interest?	Flight time (min)	Weight (lb)
DJI Mavic 2 pro	High	No	No	22	2
DJI Mavic Air 2	High	Yes	Yes	34	1.26
DJI Mavic Air	Medium	No	No	13	0.95
DJI Mavic mini	Medium	No	No	22	0.55

- A drone can autonomously conduct a mission, including auto-takeoff and auto-landing.
- The operator designs a flight path that covers the search area.
- Using images and BLE signals of the drone, the 3D map of the victims is produced.
- Drone cameras are transmitting real-time videos and images to the base station and recording high-resolution aerial video to use for post-processing.

7.4.2 Generate the 2D Map of the Victim's Position (Global Map)

The proposed architecture aims to detect survivors of disaster sites and a timely localization of the victims. These two elements are the essential motivation for the overall framework. For this experiment of localization, all the victims will have a BLE device, and the flying drone will have a BLE detector with camera Lidar. A mobile BLE detector android application will be used as a BLE detector. The flying drone will collect the data and send those data to the rescue control center to generate 2D map of the victim.

7.4.3 Generate the 3D Map Using 2D Map of Victim's Position (Local Map)

IR depth camera with Lidar data will be used as the digital 3D representations of the target. On the other hand, the BLE detector will provide the power, RSSI, and Angle-of-Arrival (AoA) for each BLE-enabled device. Using those values, one can find out the coordinate of the stuck victims with respect to the collapsed building's x-, y-, and z-axis.

7.4.3.1 Mathematical Equation and Rules for the Reachable Points of the 2D Map

- Most BLE scanners, or receivers, have an RSSI.
- The beacons transmit strength A.
- Distance $d = 10\ (A - \text{RSSI})/10n\ (n - 2$ to 4 depends on the condition).
- Angle-of-Arrival (AoA): to determine the spatial location of another Bluetooth device α.

$$x = d.\text{Sin}\ \alpha$$

Using the BLE's transmit strength, RSSI value, distance equation, and the Angle-of-Arrival (AoA) values, it is possible to easily find out the position (x, y, z) value of victims trapped inside of the collapsed building. Here below are some rules for the distance measurements:

(a) If $0 > \alpha < 90$, distance from the x-axis, $(X^{'} - x) = d$. Sin α, $X^{'}$ distance of drone from y axis
(b) If $90 > \alpha < 180$, distance from the x-axis, $(X^{'} + x) = d$. Sin α, $X^{'}$ distance of drone from y axis
(c) If $\alpha = 90°$, distance from the x-axis is X', drone height h will be the distance from the x axis, h is obtained from the drone sensor for the height measurement.

7.4.3.2 Image Analysis for 3D Mapping

Image analysis from the camera also gives us some idea about the location of the victims. When the drone is deployed inside of the collapsed building, the Lidar provides a 2D map of surrounding structures. This global map guides the maneuver of the drone by displaying possible access points. The robot then utilizes a depth camera to visualize the local 3D view. The stereoscopic infrared camera is employed to detect the survivors even in the dark. With the support of IMU, the flying drone achieves the global and local mapping of disaster scenes.

7.4.3.3 Fuzzy Logic for 3D Mapping

Generate 25 rules to get z value of the non-reachable 2D points using fuzzy model generated from the drone reachable 2D points and build 3D map of those non-reachable points. Fuzzy mapping is nonlinear that derives its output based on fuzzy reasoning and a set of fuzzy if-then rules. Mapping domain and range could be fuzzy sets/fuzzy values or points in a multidimensional spaces. We generated non-additive Mamdani fuzzy rules base to obtain output as a distance. We treat longitude as x and latitude as y and build the fuzzy rules base. Below are the rules for the Mamdani inference system:

1. If x is very small and y is very small, then z is medium.
2. If x is very small and y is small, then z is high.
3. If x is very small and y is very medium, then z is very high.
4. If x is very small and y is high, then z is very high.
5. If x is very small and y is very high, then z is very high.
6. If x is small and y is very small, then z is small.
7. If x is very small and y is small, then z is medium.
8. If x is very small and y is medium, then z is high.
9. If x is very small and y is high, then z is very high.
10. If x is very small and y is very high, then z is very high.
11. If x is medium and y is very small, then z is very small.

12. If x is medium and y is small, then z is small.
13. If x is medium and y is medium, then z is medium.
14. If x is medium and y is high, then z is high.
15. If x is medium and y is very high, then z is very high.
16. If x is high and y is very small, then z is very small.
17. If x is high and y is small, then z is very small.
18. If x is high and y is medium, then z is small.
19. If x is high and y is high, then z is medium.
20. If x is high and y is very high, then z is high.
21. If x is very high and y is very small, then z is very small.
22. If x is very high and y is very small, then z is very small.
23. If x is very high and y is medium, then z is very small.
24. If x is very high and y is high, then z is very small.
25. If x is very high and y is very high, then z is medium.

A combination of all 3D maps provides the total number of victims and their location inside of the collapsed building during search and rescue operations.

7.4.4 Cluster 3D Map Points Using K-Means Clustering Algorithms

For a single operation, Cluster 2D points of the 3D map using the K-means clustering algorithm determine the centroid of those points. The number of centroids is chosen using the Elbow Method. K-means clustering algorithms can be applied on 3D points, but for the beginning phase of the research, 2D points are used for the clustering. The rescue worker starts the rescue operation using those centroids and rescues the maximum number of victim's earliest possible time.

7.4.5 Drone Trajectory Data Mining

This step implements after completing a few searching experiments. For multiple operations of the same scenario, drone's behavior is the same because of the similar structure of the industrial building. Analyzing the drone's searching behavior of these operations can easily predict the drone's behavior in an unknown scenario. Similarly, the pattern of the victim's 3D map inside the building in an unknown scenario can be predicted by analyzing the 2D points gathered from the initial drone's BLE searching operation from the top of the collapsed building. Finally build two 3D maps for the same scenario, one 3D map using the drone's trajectory data mining and another one by physical testing. Evaluate the proposed architecture by comparing these two 3D maps.

At the end of the research, find the accuracy of the proposed architecture by comparing the predicted result with experimental result.

7.5 Results and Analysis

Using the proposed architecture, we tested the drone system in a confined space. First, we scanned the top of the building using the drone's BLE scanner and found out the victim's position points. Then we filtered those points using different techniques and found out the points to start the rescue operation using the data mining approach. Here below are the different steps of the experiment.

7.5.1 Accessibility of the Points

Few points are visually accessible by drone and few are not. For visually accessible points, calculate the 3D values from 2D values using the proposed mathematical equation.

$$x = d.\mathrm{Sin}\ \alpha, d = 10\ (A - \mathrm{RSSI})\ /10n \quad (n = 2 \text{ for the existing condition})$$

Next build a Fuzzy Inference System from visually accessible points and implemented on those visually non-accessible points and calculate the 3D values of those points. The results are illustrated in Fig. 7.2, FIS using Math lab.

After filtering those points from step 5.1, we used those points for K-means clustering and found the centroid for the rescue operation for rescue workers using 2D points. At the same time, we added the 3D points obtained from step 5.1 to get the final 3D map and the overall situation of the centroid's depth of the rescue operation. Section 7.5.3 provides the K-means clustering result, using the Weka Machine learning workbench.

7.5.2 K-Means Clustering

To find out the feasible center points of the operation to rescue the victims with minimum time and maximum accuracy (Fig. 7.3).

From the above cluster analysis, we got two centroids for the rescue operation. Table 7.2 shows the outcome of the rescue operation. In about 30 min, we can find out 7 victims out of 10 victims.

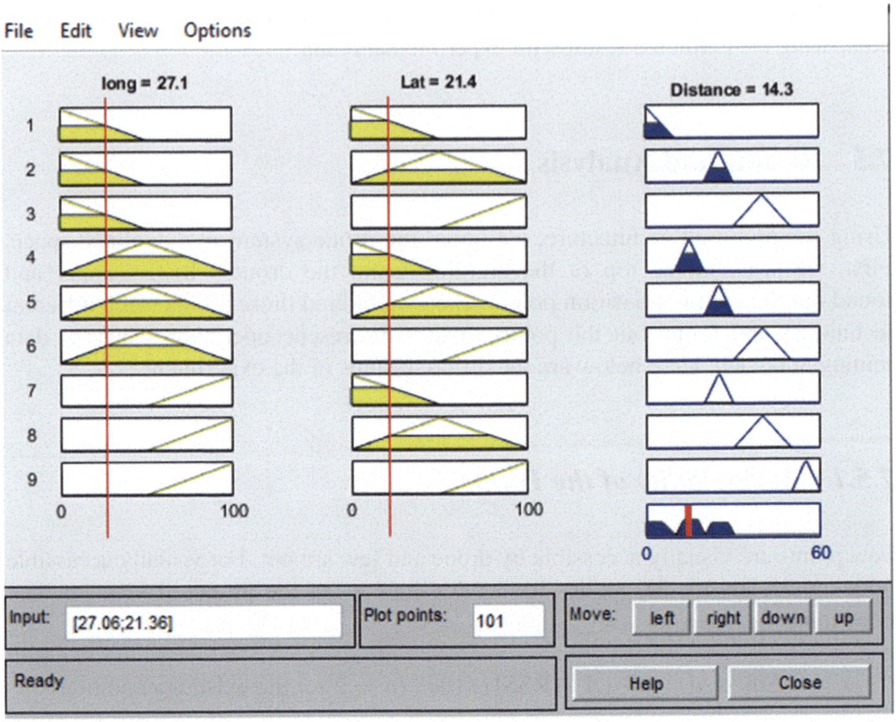

Fig. 7.2 Calculate 3D value of non-accessible points using FIS

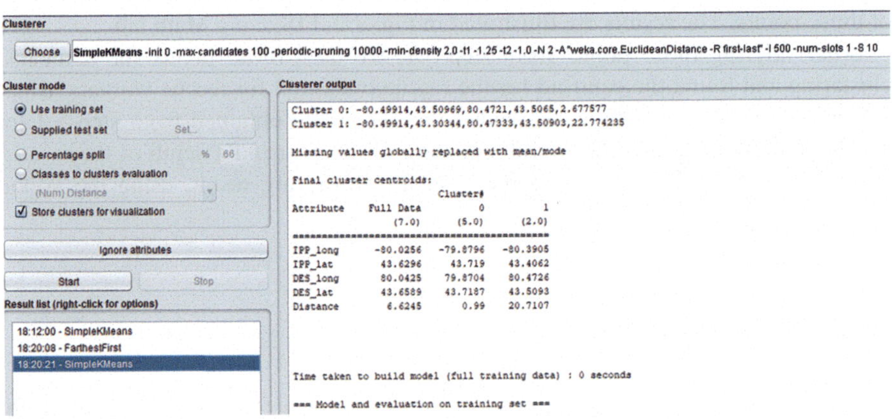

Fig. 7.3 K-means clustering of the acceptable points of the victims

Table 7.2 Outcome of the rescue operation (training accuracy)

Scenario no.	Total 2D points	Number of cluster	No. of victims	No. of victims rescued	Time to rescue the victims (min)
1	8	2	10	7	30

Table 7.3 Outcome of the rescue operation (test accuracy)

Scenario no.	No. of victims present	No. of victims rescued
1	10	6

7.5.3 *Drone Trajectory Data Mining: Drone-Led Search Efforts in Unknown Situation*

Finally, in an unknown situation, we did our drone scanning operation from the top of the building and predict the victim's position points inside of the building from previous experiments. Table 7.3 shows the outcome of the rescue operation using drone trajectories.

7.5.4 *Accuracy Measurement*

From the model based on the BLE scanning result, number of victims present inside of the building is A, and from the physical experiment of the same scenario, the number of victims present inside of the industrial collapsed building is B. So the accuracy of the model is $(A/B)*100$. In our case, $A = 6$, $B = 10$, so the accuracy of our proposed system is 60.

7.6 Conclusion

This study provides an overall architecture of the drone assist disaster management system and presents suitable drone hardware with sensors and software assist detector for practical search and rescue operations. This system was tested in wall-covered, low-light, GPS-denied conditions and visualized the position of victims using the test data. Furthermore, the system generated a model using fuzzy logic, k-means clustering, and drone trajectory data mining for the searching operation and rescue operation. This model is tested in an unknown environment and could help rescuers to rescue victims from the industrial collapsed building at the shortest time.

It is worthwhile looking into heartbeat detector mobile applications as further upgrades to the drone architecture.

References

1. Lee, S., Har, D., & Kum, D. (2016). Drone-assisted disaster management: Finding victims via infrared camera and lidar sensor fusion. In *3rd Asia-Pacific world congress on computer science and engineering*.
2. Sehrawat, A., Choudhury, T., & Raj, G. (2017). Surveillance drone for disaster management and military security. In *International conference on computing, communication and automation (ICCCA)*.
3. Dinh, T. D., Pirmagomedov, R., Pham, V. D., Ahmed, A. A., Kirichek, R., Glushakov, R., & Vladyko, A. (2019). Unmanned aerial system–assisted wilderness search and rescue mission. *International Journal of Distributed Sensor Networks, 15*(5), 10.
4. Grzonka, S., Grisetti, G., & Burgard, W. (2009). Towards a navigation system for autonomous indoor flying. In *IEEE international conference on robotics and automation (ICRA)*.
5. Bachrach, A., He, R., & Roy, N. (2009). Autonomous flight in unstructured and unknown indoor environments. In *Proceedings of EMAV*.
6. Engel, J., Sturm, J., & Cremers, D. (2012). Camera-based navigation of a low-cost quadrocopter. In *IEEE/RSJ international conference on intelligent robots and systems (IROS)*.
7. Engel, J., Sturm, J., & Cremers, D. (2014). Scale-aware navigation of a lowcost quadrocopter with a monocular camera. *Robotics and Autonomous Systems, 62*(11), 1646–1656.
8. Hausman, K., Weiss, S., Brockers, R., Matthies, L., & Sukhatme, G. S. (2016). Self-calibrating multi-sensor fusion with probabilistic measurement validation for seamless sensor switching on a UAV. In *IEEE international conference on robotics and automation (ICRA)*.
9. Ling, Y., Liu, T., & Shen, S. (2016). Aggressive quadrotor flight using dense visual-inertial fusion. In *Ieee international conference on robotics and automation (ICRA)*.
10. Chen, J., Liu, T., & Shen, S. (2016). Online generation of collision-free trajectories for quadrotor flight in unknown cluttered environments. In *IEEE international conference on robotics and automation (ICRA)*.
11. López, E., et al. (2015). Indoor SLAM for micro aerial vehicles using visual and laser sensor fusion. In *Second Iberian robotics conference* (pp. 531–542).
12. Saska, M., et al. (2016). System for deployment of groups of unmanned micro aerial vehicles in GPS-denied environments using onboard visual relative localization. *Autonomous Robots, 41*, 1–26.
13. Naghdi, S., & O'Keefe, K. (2020). Detecting and correcting for human obstacles in BLE trilateration using artificial intelligence. *Sensors, 20*(5), 1350.

Chapter 8
Optimizing the Key-Pair Generation Phase of McEliece Cryptosystem

Michela Ceria, Alessandro De Piccoli, Martino Tiziani, and Andrea Visconti

8.1 Introduction

Quantum computing theory and even more quantum computers represent a threat for public key cryptography. Nowadays, they have been built but they do not still have enough qbits to actually break a cipher like RSA implemented with the recommended bit security level. In addition, key establishment schemes and digital signature algorithms are not secure [12, 13] anymore. Therefore, in the last years, researchers have suggested new cryptographic algorithms which are resistant to quantum and classic computers, and we will refer to them as Post-Quantum Cryptography (PQC).

The National Institute of Standards and Technology (NIST) published a call for proposals in order to find new good standards for Post-Quantum Cryptography. Currently, we reached the third round, with seven finalists: four Public-key Encryption and Key-establishment Algorithms, namely Classic McEliece (code-based), CRYSTALS-KYBER (Learning With Errors), NTRU (lattice-based), SABER (Module Learning With Rounding) and three Digital Signature Algorithms, namely CRYSTALS-DILITHIUM (lattice-based), FALCON (lattice-based) and Rainbow (multivariate). Moreover, there are eight Alternate Candidates, divided as follows: five Public-key Encryption and Key-establishment Algorithms called

M. Ceria
Politecnico di Bari, Department of Mechanics, Mathematics and Management (DMMM), Bari, BA, Italy

A. De Piccoli · M. Tiziani · A. Visconti (✉)
Università degli Studi di Milano, Department of Computer Science "Giovanni Degli Antoni", Milano, MI, Italy
e-mail: alessandro.depiccoli@unimi.it; martino.tiziani@studenti.unimi.it; andrea.visconti@unimi.it

I. Woungang, S. K. Dhurandher, *4th International Conference on Wireless, Intelligent and Distributed Environment for Communication*, Lecture Notes on Data Engineering and Communications Technologies 94, https://doi.org/10.1007/978-3-030-89776-5_8

BIKE (code-based), FrodoKEM (lattice-based), HQC (code-based), NTRU Prime (lattice-based), SIKE (supersingular curves isogeny) and three Digital Signature Algorithms, i.e. GeMSS (multivariate), Picnic (zero-knowledge), SPHINCS+ (hash-based). Despite they were well known, the post-quantum cryptographic algorithms had to address a number of issue [17]. For example, the size of the keys was not negligible, and post-quantum algorithms were not optimized for real-time applications and resource-constrained devices.

In this paper we will focus on **Classic McEliece**, a promising algorithm that reached the third round of the NIST Post-Quantum Cryptography Standardization Process. This algorithm is based upon a public key cryptosystem by Robert J. McEliece [15] but has to address the problem of its large key sizes. Notice that the key generation process takes a non-negligible time if compared with the encryption/decryption process. One may think that there is no need to speed up the key generation but, for instance, think about generating millions of keys for the citizens of some state. In such a use case, a simple optimization may positively affect the whole key generation process. When generating public and private keys of Classic McEliece cipher, the most important operation is polynomial multiplication. In recent years, a number of papers addressed the problem to speed up the polynomial multiplication, in particular in characteristic 2 fields [3, 6–9, 14, 18, 21] using techniques from [5, 16]. Our aim is to exploit such speedups (but not only) to improve the performances of the key-pair generation phase.

The paper is organized as follows. In Sect. 8.2, we summarize the McEliece algorithm. In Sect. 8.3, we explain where optimizations take place. In Sects. 8.4 and 8.5, we analyse the details of speedups for private and public key, respectively. Finally, in Sect. 8.6, we present a detailed report about the testing results. Some conclusions are then stated in Sect. 8.7.

8.2 Classic McEliece

In this section, we briefly recall the code-based [1] Classic McEliece cryptosystem that we will optimize in what follows.

Let n be the length of the code involved in our cryptosystem and $t \geq 2$ its error-correction capability. We consider a finite field $\mathbb{F}_q \equiv \mathbb{F}_2[z]/f(z)$, where $q = 2^m$ and $m \in \mathbb{N}$, such that $n \leq q$, $mt < n$ and $f(z)$ is a monic irreducible polynomial in $\mathbb{F}_2[z]$ of degree m, defining a representation of the finite field. This defines the code dimension $k = n - mt$.

We see now how the keys are generated:

1. Select $g(x) \in \mathbb{F}_q[x]$, a random monic irreducible polynomial.
2. Select $(\alpha_1, \alpha_2, \ldots, \alpha_n)$, n random distinct elements of $\mathbb{F}_q$.
3. Compute the $t \times n$ matrix $\hat{H}$, where $\hat{H}_{ij} = \dfrac{\alpha_j^{i-1}}{g(\alpha_j)}$, for $i = 1, \ldots, t$ and $j = 1, \ldots, n$.

4. Compute the $mt \times n$ matrix H, replacing $\hat{H}_{ij} = a_0 + a_1 x + a_2 x^2 + \dots$ with its binary representation $a_0, a_1, a_2, \dots$.
5. Reduce H in the standard form $H = [\mathbb{I}_{n-k}|T]$, through a binary Gaussian elimination; if it is not possible return to the first step.
6. The $(n-k) \times k$ matrix T is the public key, while $(g(x), \alpha_1, \alpha_2, \dots, \alpha_n)$ is the private key.

We move now to the encryption phase:

1. Select a binary message e, which is a column vector of length n and weight t [19].
2. Construct $H = [\mathbb{I}_{n-k}|T]$ using the public key.
3. Compute the ciphertext $c = He$.

Finally, we show the decryption procedure:

1. Define $v = (c, 0)$, appending k zeroes to c.
2. Compute the syndrome $s = Hv$ and transform it in a polynomial of $\mathbb{F}_q[x]$.
3. Find the value e, using s as input in the Berlekamp–Massey algorithm; if $w(e) = t$ and $c = He$, e is the right message, otherwise the algorithm failed.

8.3 Software Optimization

In what follows, we present new software optimizations that make use of the following technologies: `CLMUL` set instruction extension [10, 20], advanced polynomial multiplication and a parallel implementation for the public key generation.

Classic McEliece [4] has been proposed as a candidate to the NIST Post-Quantum Cryptography with four different software implementations; due to its better performances, the AVX implementation will be used as basis for the optimizations we will develop in what follows.

Experimental results show that the big bottleneck of this cipher is the *key-pair generation*. This is due to the high dimension of the public key which, in this parameters' configuration, reaches almost 1.4MB: compared to a 2048 bit RSA key, it is clear why this cipher is not recommended in a lightweight environment. On the other hand, the encapsulation and decapsulation phases appear to be very efficient, providing execution times that are *in average* lower than a millisecond. Due to the previous insight, we concentrate our improvements only on the key-pair generation. We identified in this phase *five different operations* that need strong optimization, in order to obtain a global speedup: polynomial multiplication (private key), Gaussian reduction (private key), Gaussian reduction (public key), computation of a linear map (public key) and application of the linear map (public key).

8.4 The Private Key

In McEliece cryptosystem the private key is a Goppa code; in *mceliece8192128* this code is represented by an irreducible Goppa polynomial of degree $t = 128$, obtained by a Gaussian reduction of a randomly defined 129×128 matrix, initialized by the multiplication of two elements in $GF(2^{13})^{128}$, expressed in polynomial form.

We start by optimizing the polynomial multiplication in $GF(2^{13})^{128}$, which relies on the multiplication of their coefficients in $GF(2^{13})$, represented again as polynomials over $GF(2)$. This means that we have to optimize two different multiplications, namely, `gf_mul_13` for $GF(2^{13})$ and `gf_mul_13_128` for $GF(2^{13})^{128}$. The `gf_mul_13` multiplication takes as an input two polynomials that can be represented in software in a very efficient way as 16-bit integer values. The speedup has been obtained via the `CLMUL` and the `PCLMULQQD` instructions [10, 20]. We can perform the operation in 5 instructions instead of the almost 50 required by the original algorithm.

The `gf_mul_13_128` multiplication requires a much more complex optimization, involving the usage of advanced multiplication algorithms as the one described in [8]. This algorithm, applied to the *mceliece8192128* parameters, generates a total of 11,309 operations, 3888 of which are multiplications and 7420 summations.

For a software implementation a problem needs to be addressed: The intermediate results, produced during the computation, need to be stored in memory and retrieved to calculate the subsequent intermediate results. Since the multiplication is repeated 127 times, we get some advantages in environment with cache memory.

The needed modular reduction on the 25-bit polynomials we get for `gf_mul_13` cannot be improved, but since we store all 11,309 intermediate results of the products, we can avoid to reduce at all: The result of a multiplication on $GF(2^{13})$ will never be the input of another multiplication but only of summations, never growing over 25 bits. If we double the memory requirements for memorizing the intermediate results, the modular reduction of `gf_mul_13_128` can be almost completely removed. It remains necessary only for the final 255 polynomials that represent the coefficients of the `gf_mul_13_128`'s result. After initializing the matrix with the polynomial multiplication, Classic McEliece reduces it by means of the Gaussian elimination. Our optimization is based on two aspects, namely that the reduction's algorithm can be substituted with a better one like PLU, PLUQ or CUP decomposition [11] and that the reduction's basic operations can be optimized (therefore, we are again using our multiplication optimization). For the first aspect, the M4RIE [2] library is the perfect tool. It offers three different Gaussian reductions: naive, ple, newton-john and, among them, the one performing better for small matrices is *naive*. The algorithm used for the multiplication in M4RIE is very inefficient compared to `gf_mul_13`, so we substituted it with `gf_mul_13`. The combination of the two techniques generates a huge speedup.

8.5 The Public Key

The data structures used in the public key's creation are two matrices. For *mceliece8192128* the main one is a 1664×128 matrix (`mat`) and the second one is a 1664×26 (`ops`), both of which hold 64-bit integer values. The algorithms we want to optimize share a basic logic in common: Every operation performed on the data structures is a mask application, obtained with a logical `AND` operation (&) between one element of the matrix and one temporary element (the mask). Another common feature is the low dependency between operations and data; this means that we can produce an efficient parallel implementation of the same algorithms. These common features allow a simple and similar optimization on all of the targeted operations and, with a good memory handling, these improvements can grant a huge speedup in the computation of the public key.

Changing the implementation with a parallel one may change the memory access pattern of the algorithm; rearranging rows and columns can drive to very different performances. In order to obtain a good speedup, we need to modify the matrices' memory layout to allow efficient access in their parallel implementation. Gaussian reduction and linear map computation use only the first 26 columns of `mat` and the entire `ops`. The row access on these matrices is efficient, the one in columns is not. The application of the linear map uses the last 102 columns of `mat` and the entire `ops`. Both row and column accesses to `mat` are efficient because now the number of columns is not as small as before.

The Gaussian elimination algorithm performs an iterative reduction of a sub-matrix in order to bring the initial matrix into echelon form. The sub-matrix reduction is a complete parallel task (100% capable, without dependencies among operations and data), which can be implemented efficiently with multiple threads computation. The synchronization step slows down the parallel computation, but this is unavoidable.

All the masks' applications depend entirely on one single column i. If this column is computed before the other 25 ones and the intermediate masks are saved in temporary memory, we can apply the masks to the other 25 columns completely in parallel without any synchronization among the columns. This opens up to a good parallel implementation that heavily differs from the classical parallel Gaussian implementation. Every thread accesses a portion of the 25 columns and independently applies the masks to the matrix. To have an efficient access to the columns we apply a transposition to the matrices. For the implementation, we choose the OpenMP since it offers the thread-pool concept and allows a fast and easy usage. We opt for a parallel section that encloses the entire algorithm, every thread works only on a subset of columns (the dimension is determined at runtime) and the threads synchronize on each, outermost, iteration by a `#pragma omp barrier`. The column, i, that generates dependency among the other columns, needs to be computed in a non-parallel section and this can be realized with the `#pragma omp single` directive. The masks' applications work in parallel where every thread applies the masks to its columns subset. In the linear map's creation the mask is based on a read-only data and does not depend on previous values; therefore, the application can be executed on all 26 columns without any synchronization construct. The implementation follows that for the Gaussian reduction, the entire algorithm is enclosed with a parallel region, every thread processes a subset of columns, but this time no synchronization is required, allowing the parallel computation to run at full speed. The last step of the public key creation is the linear map's application to the matrix `mat`.

The map application's complexity is higher than any other operation in the cipher and, thanks to the 100% parallel implementation, we expect a great performance boost. The parallelization is more or less the same as before, a unique parallel region that encloses the entire algorithm, with no synchronization because the mask depends on read-only data. The main problem is the temporary array that we need to store (`one_row`). In a sequential implementation we have only one array but, in parallel, we need one array for each thread. The memory cost of defining a new array for every thread is too high and so it is not affordable. A smarter approach is to use the first 26 columns of `mat` as a buffer to hold the intermediate values of `one_row`.

The memory poses another problem. At this point `ops` is not in the original form, but in its transposed one (`ops_t`). However, the original algorithm performs access on `ops`'s rows, which are inefficient on `ops_t`. To cope with this issue we have to transpose again the rows to their original form. In order to do that, we transpose `ops_t`'columns onto `mat_t`'rows. Thanks to this unusual transformation all memory accesses become efficient and the parallel implementation can run at full speed.

8.6 Testing and Results

In this section, we test our improvements to see to which extent we can speed up the original implementation. The test machine is a Dell XPS 15 9560, with the following specifications:

- CPU: Intel i7-7700HQ, 2.8GHz to 3.8GHz, 4 Core.
- RAM: 8GB 2400MHz.
- GPU: Nvidia 1050 (not used by the cipher).
- OS: Ubuntu 19.10 virtualized with VMWare on a Windows 10 host system.

All tests have been conducted with the better performance mode offered by the machine, using the electrical grid instead of the battery, as the power source. The test of a specific optimization has been repeated many times, to compensate variations on the machine status.

As illustrated by the specifications of the XPS machine, the CPU has four physical cores and eight virtual threads, respectively. As stated before, the parallelism can be a great tool, but it requires a very careful setup. In theory, we could run the parallel optimizations with eight different logical threads (as the CPU allows), but virtual threads share the same physical cores. This means that the threads will be in conflict for the same hardware resources, thing that, in cryptography, may become quite a big problem. In this scenario, the waiting imposed to the threads would dramatically slow down the computation, instead of speeding it up. With this in mind, the better configuration is to use the same amount of logical threads as the number of physical cores the CPU contains.

One can improve even further with the binding of every logical thread to a specific core: During the entire computation the thread will never switch to another core, reducing the management overhead. Thanks to these considerations, we have conducted all the following tests with four logical threads instead of eight. This also means that the obtained results can be further improved with a more powerful CPU. The optimization's performances are highly bounded to the hardware and configuration uses for the computation, a good mix of the two can deliver a great speedup as reported in the following test.

8.6.1 Private Key

8.6.1.1 Polynomial Multiplication

This operation has been optimized in its two components `gf_mul_13` and `gf_mul_13_128`. Thanks to `PCLMULQQD`, we can remove almost 40 machine instructions from `gf_mul_13`; this translates to a very fast multiplication, which allows a good speedup on the private key computation since `gf_mul_13` is used also in the Gaussian reduction. To evaluate this optimization, we perform 10 million multiplications and we report the results in Table 8.1. The result of this test gives an idea on how much we gain from this dedicated, machine instruction, which directly executes a complex task. We have a reduction in the computation times of almost an order of magnitude.

The `gf_mul_13_128`'s optimization gives a speedup of approximately 60%. The following test concerns the timing of the entire matrix initialization phase. This means that a single test repeats the `gf_mul_13_128` operation 127 times. Table 8.2 reports the performance of 4000 consecutive test.

8.6.1.2 Gaussian Reduction

Thanks to the very fast polynomial multiplication, the advanced reduction algorithm can achieve its maximum performance, with a speedup of almost 75%. This

Table 8.1 Computation times of `gf_mul_13` over 10,000,000 tests

`gf_mul_13`	Original	Optimized
Total time (ms)	84,356	3,634

Table 8.2 Performance comparison of the entire polynomial multiplication phase over 4000 tests

Percentile	Original (ms)	Optimized (ms)	Speedup (%)
25%	3,149	1,271	59.63
50%	3,211	1,281	60.00
75%	3,415	1,299	61.96

Table 8.3 Private key Gaussian reduction comparison over 4000 tests

Percentile	Original (ms)	Optimized (ms)	Speedup (%)
25%	8,201	2,303	71.91
50%	8,320	2,313	72.20
75%	9,005	2,334	74.00

Table 8.4 Public key Gaussian reduction comparison, over 4000 tests

Percentile	Original (ms)	Optimized (ms)	Speedup (%)
25%	37,329	28,151	24.58
50%	39,898	28,586	28.35
75%	43,073	35,103	18.50

represents a huge increment of performance for the private key generation since the Gaussian reduction constitutes almost 67% of its computation time. Table 8.3 reports the results of 4000 tests, where the same matrix has been reduced with the two different implementations.

8.6.2 *Public Key*

The public key's test can be negatively affected by the machine status and configuration. Therefore, we conducted the following tests in the most stable and quiet environment possible, with no other applications running (other than the related OS).

8.6.2.1 Gaussian Reduction

This parallel optimization is particularly difficult because of all the synchronization required to correctly execute the elimination algorithm; also, for the same reason, the performance speedup is not as relevant as the one obtained on the optimizations of the map creation and application. To allow a fast execution of this optimization, a change in the memory layout may be necessary. In Table 8.4, the timings of the two versions are summarized; it is easy to see that this operation is very complex to optimize and even with a parallel implementation, the speedup barely reaches the 28%. The problem related to the machine status can be seen in the maximum

Table 8.5 Generation of the linear map, over 4000 tests

Percentile	Original (ms)	Optimized (ms)	Speedup (%)
25%	10,877	2,380	78.00
50%	11,117	2,398	78.42
75%	12,092	2,468	79.60

Table 8.6 Application of the linear map, over 4000 tests

Percentile	Original (ms)	Optimized (ms)	Speedup (%)
25%	62,112	17,474	71.86
50%	63,216	17,624	72.12
75%	70,300	18,574	73.57

computation time registered, with 90,698 ms for the original implementation and 115,134 ms for the optimized one.

8.6.2.2 Linear Map Creation

Thanks to the memory layout modifications, this operation can be executed 100% in parallel without any synchronization, thus producing very good performance. With these proprieties, the speedup is directly related with the number of cores/threads at disposal and having zero synchronization means to have a close to zero management overhead, so the threads can run at full speed producing good overall performance. Table 8.5 reports the timings of 4000 tests; the results indicate that this is the best optimization in terms of percentage increment, with a peak of almost 80% speedup.

8.6.2.3 Application of the Linear Map

The same considerations made for the previous operation can be also applied to this one: Full parallel implementation with no synchronization is required. As reported in Table 8.6, the test indicates a lower percentage increment, but, since this operation is by far more complex and costly than all the others, the overall decrease of computation time is definitely the best one. Thanks to this optimization, the entire key-pair generation highly reduces its impact on the global execution time.

8.6.3 Key-Pair

We give a global review of the two implementations' timings, indicating the overall optimization's performance and speedups. In Table 8.7 we reported the timings of 2000 sequential tests. The results indicate a global speedup of 55% on the fourth quartile, and this represents quite a good improvement: The execution times are almost halved.

Table 8.7 Timings compare, over 2000 tests

Percentile	Original (ms)	Optimized (ms)	Speedup (%)
Private key generation			
25%	12, 201	4, 599	62.30
50%	12, 356	4, 628	62.50
75%	13, 421	4, 675	65.16
Public key generation			
25%	111, 852	49, 388	55.84
50%	115, 142	51, 385	55.37
75%	127, 750	67, 702	47.00
Key-pair generation			
25%	185, 268	98, 963	45.70
50%	191, 184	107, 776	43.63
75%	212, 542	118, 116	55.57

8.6.4 Memory Consumption

In order to have a good optimization, the additional memory requirements need to be very low, to prevent the explosion of an already problematic parameter. In this section, we give a brief overview of the memory consumption of every defined optimization.

- **Polynomial multiplication**: The `gf_mul_13`'s optimization uses only the `PCLMULQDQ` instruction, so no additional memory is required; `gf_mul_13_128` requires the memorization of the entire intermediate results produced by the multiplication technique. In this configuration the intermediate results are 11,309 25-bit polynomials (remember the removal of the modulo reduction); every polynomial is represented with a 32-bit variable, meaning a total memory consumption of about 44 KB.
- **Private-key Gaussian reduction**: This optimization is based on the `naive reduction` function offered by the `M4RIE` library. Since the memory requirement for this optimization depends entirely on a third party library, we do not include this quantity into the total count, because it could change basing on the library's version or future modification.
- **Public-key Gaussian reduction**: This optimization requires the memorization of two different columns (holding the masks which generate the dependencies) from the `mat` matrix. The matrix has 1664 row and every element is a 64-bit variable; therefore, the additional memory required is 25 KB.
- **Computation of the linear map**: No additional memory is required.
- **Application of the linear map**: Thanks to the smart reuse of `ops_t` no further memory is required.

The analysis shows that the total additional memory used by the optimizations is, approximately, 70 KB, while the total amount of memory required by *mceliece8192128* is 3.3 MB (1.3 MB for the public key memorization; + 2MB for the matrix used to compute the public key). The additional memory required by the optimizations (excluding the `M4RIE`) is the 2% of the global memory required by Classic McEliece. At the end, the optimization's evaluation is very good: Using almost 2% additional memory, we can halve the execution time with a consumer CPU.

8.7 Conclusions

After recalling the Classic McEliece code-based cryptosystem, in this paper we suggest how to speed up the key-pair generation phase. Such a phase is a bottleneck for this cipher; indeed, the public key is large and this may negatively affect the use of this cipher in specific environments. More precisely, we optimized a number of operations: polynomial multiplication (private key), Gaussian reduction (private and public keys), computation of a linear map (public key) and application of the linear map (public key). In order to measure the optimizations suggested, we executed an intensive testing activity, showing that, with a small amount of additional memory used, the execution time has been almost halved.

References

1. Adámek, J.: *Foundations of coding - theory and applications of error-correcting codes with an introduction to cryptography and information theory*. Wiley (1991)
2. Albrecht M. R. (2012). The M4RIE library for dense linear algebra over small fields with even characteristic (pp. 28–34). http://doi.acm.org/10.1145/2442829.2442838
3. Bernstein, D. J. (2009). Batch binary Edwards. In S. Halevi (Ed.), Advances in Cryptology - CRYPTO 2009 (pp. 317–336). Springer Berlin Heidelberg.
4. Bernstein, D. J., Lange, T., von Maurich, I., Misoczki, R., Niederhagen, R., Peters, C., Schwabe, P., Szefer, J., Wang, W., Chou, T., & Persichetti, E. (2020). *Classic McEliece: conservative code-based cryptography*. NIST, National Institute of Standards and Technology. https://csrc.nist.gov/projects/post-quantum-cryptography/round-3-submissions
5. Boyar, J., & Peralta, R. (2010). A new combinational logic minimization technique with applications to cryptology. In P. Festa (Ed.) *Experimental Algorithms* (pp. 178–189). Springer Berlin Heidelberg.
6. Cenk, M., & Hasan, M. A. (2015). Some new results on binary polynomial multiplication. *Journal of Cryptographic Engineering*, *5*(4), 289–303. https://doi.org/10.1007/s13389-015-0101-6
7. Cmt: Circuit minimization work. http://www.cs.yale.edu/homes/peralta/CircuitStuff/CMT.html
8. De Piccoli, A., Visconti, A., & Rizzo, O. G. (2019) Polynomial multiplication over binary finite fields: new upper bounds. *Journal of Cryptographic Engineering*. https://doi.org/10.1007/s13389-019-00210-w
9. Find, M. G., & Peralta, R.: Better circuits for binary polynomial multiplication. *IEEE Transactions on Computers*, *68*(4), 624–630 (2019). https://doi.org/10.1109/TC.2018.2874662

10. Jankowski, K., & Pierre Laurent, A. O. (2012). *Intel polynomial multiplication instruction and its usage for elliptic curve cryptography*. Tech. rep., Intel.
11. Jeannerod, C. P., Pernet, C., & Storjohann, A. (2013). Rank-profile revealing Gaussian elimination and the cup matrix decomposition. *Journal of Symbolic Computation*, *56*, 46–68. https://doi.org/10.1016/j.jsc.2013.04.004. http://www.sciencedirect.com/science/article/pii/S0747717113000631
12. Ma, L., Song, J., Whitenton, E., Zheng, A., Vorburger, T., & Zhou, J. (2004). NIST bullet signature measurement system for RM (reference material) 8240 standard bullets. *Journal of Forensic Science*, *49*(4), 1–11. https://doi.org/10.1520/JFS2003384
13. Mavroeidis, V., Vishi, K., Zych, M. D., & Jøsang, A. (2018). The impact of quantum computing on present cryptography. *International Journal of Advanced Computer Science and Applications*, *9*(3). https://doi.org/10.14569/ijacsa.2018.090354
14. Maximov, A., & Ekdahl, P. (2019). New circuit minimization techniques for smaller and faster AES Sboxes. *IACR Transactions on Cryptographic Hardware and Embedded Systems*, *2019*, 91–125. https://doi.org/10.13154/tches.v2019.i4.91-125. https://tches.iacr.org/index.php/TCHES/article/view/8346
15. McEliece, R. J. (1978). A public key cryptosystem based on algebraic coding theory. The Deep Space Network Progress Report 42–44 (pp. 114–116), National Aeronautics and Space Administration.
16. Paar, C., & Rosner, M. (1997). Comparison of arithmetic architectures for Reed-Solomon decoders in reconfigurable hardware. In *Proceedings. The 5th Annual IEEE Symposium on Field-Programmable Custom Computing Machines Cat. No. 97TB100186* (pp. 219–225).
17. Perlner, R. A., & Cooper, D. A. (2009). Quantum resistant public key cryptography: A survey. In *Proceedings of the 8th Symposium on Identity and Trust on the Internet, IDtrust '09* (pp. 85–93). Association for Computing Machinery. https://doi.org/10.1145/1527017.1527028
18. Reyhani-Masoleh, A., Taha, M. M. I., & Ashmawy, D. (2018). Smashing the implementation records of AES S-box. *Transactions on Cryptographic Hardware and Embedded Systems*, *2018*(2), 298–336. https://doi.org/10.13154/tches.v2018.i2.298-336. https://tches.iacr.org/index.php/TCHES/article/view/884
19. Sendrier, N. (2005). Encoding information into constant weight words. In *Proceedings. International Symposium on Information Theory, 2005. ISIT 2005* (pp. 435–438). https://doi.org/10.1109/ISIT.2005.1523371
20. Shay Gueron, M. E. K. (2014). Intel carry-less multiplication instruction and its usage for computing the GCM mode. Tech. rep.
21. Visconti, A., Schiavo, C. V., & Peralta, R. (2017). Improved upper bounds for the expected circuit complexity of dense systems of linear equations over GF(2). Cryptology ePrint Archive, Report 2017/194. https://eprint.iacr.org/2017/194

Chapter 9
Dual Parameter Ranking Based Resource Allocation for PD-SCMA Cognitive Radio Networks

Simon Chege and Tom Walingo

9.1 Introduction

Modern 5G wireless networks confront increased demand to support very diverse traffics with much tighter requirements of massive connectivity, spectrum efficiency, enhanced throughput and lower latency. Non-orthogonal multiple access (NOMA) technology provides a paradigm shift from orthogonal multiple access techniques (OMA), promising efficient utilization of spectrum resources in 5G wireless networks. Unlike OMA that allows only one user to use a resource element (RE), orthogonal frequency division multiple access (OFDMA) subcarrier, time division multiple access (TDMA) timeslot or code division multiple access (CDMA)'s code, NOMA supports spectrum sharing among multiple users within one resource block by exploiting power domain multiplexing, code domain multiplexing or a fusion of both schemes, resulting in high spectral efficiency, user fairness and massive connectivity [1]. NOMA schemes permit controllable interference by non-orthogonal resource allocation. The interference and RE limitation constraints necessitate the development of new optimal radio resource allocation (RRA) and user pairing schemes for these networks.

Cognitive radio (CR), in the recent past, has been envisioned to provide enhanced and intelligent wireless spectrum efficiency. CR is based on principles of spectrum sensing and dynamic spectrum access, where secondary users (SUs) sense the spectrum occupied by primary users (PUs), and intelligently adapt their operating parameters to access a spectrum band in an opportunistic or collaborative manner

---

S. Chege (✉) · T. Walingo
University of KwaZulu Natal, Durban, South Africa
e-mail: ChegeS@ukzn.ac.za; Walingo@ukzn.ac.za

I. Woungang, S. K. Dhurandher, *4th International Conference on Wireless, Intelligent and Distributed Environment for Communication*, Lecture Notes on Data Engineering and Communications Technologies 94, https://doi.org/10.1007/978-3-030-89776-5_9

while keeping the interference temperature limit within the threshold [2]. To date, research on CR technology exploitation to support emerging applications has received significant attention, where new architectures for CR networks based on full-duplex, device-to-device, and multiple-input multiple-output (MIMO) have been studied to further increase spectrum efficiency [3]. Particularly, blending of NOMA and CR promises to meet the spectral, connectivity, latency and throughput requirements of 5G networks is an active research. However, since both NOMA and CR are interference limited, where coexistence of cross-tier interference between the primary and secondary networks and co-tier interference caused by NOMA exist, practical realization of such a system is ambitious. In this article, we employ a hybrid NOMA technique in an underlay cognitive radio network where concurrent primary and secondary transmissions are allowed, under the condition that the interference on the primary network is below a controllable level. We then propose a dual parameter ranking resource allocation (DPR-RA) scheme for codebook assignment, user pairing, and power allocation for sum-rate maximization.

9.1.1 Related Work

Generally, in the last decade, NOMA schemes have received significant attention as a potential multiple access scheme for 5G technologies. Proposed in [4] is a PD-NOMA scheme that clusters the users based on their channel quality differences and performs optimal power allocation that maximizes the sum-throughput of all users. In [5], different SCMA configurations are employed to explore performance aspects for edge IOT systems. A scenario where the network operator shares the uplink spectrum between the original long-term evolution (LTE) users and new users using SCMA is investigated in [6]. In [7], SCMA scheme is proposed and a comparison made with the PD-NOMA scheme where results indicate improved sum-rate performance at the expense of increased complexity. For heterogeneous multi-tier network (HetNet) consisting of small cell user equipment's (SUEs) and the macro user equipment's (MUEs), authors in [8–10] propose a novel NOMA scheme integrating both PD-NOMA and SCMA to yield a hybrid power domain sparse code non-orthogonal multiple access (PD-SCMA) scheme. In hybrid PD-SCMA, MUEs and SUEs are co-multiplexed, using PD-NOMA, into a codebook, while allocating different power levels and hence applying successive interference cancellation (SIC) for their detection at the receiver. Furthermore, different codes are exclusively assigned to the MUE-SUE clusters hence employing a modified log-domain message passing algorithm (log-MPA) for interference cancellation.

A major objective of CR is to realize dynamic spectrum access/sharing by learning its surrounding environments and adapting its operating parameters. Currently, there exist three CR paradigms [3]: Interweave, in which a SU can transmit only when no PU occupies the licensed spectrum; Underlay, where concurrent primary and secondary transmissions are allowed, under the condition that the interference on the primary network is below a controllable level; and Overlay in which a SU provides relaying services to the primary network, and at the same time, transmits its own signal. The combination of NOMA and CR for 5G requirements is investigated in [11–13]. In [11], the authors investigate the application of NOMA to a cooperative spectrum-sharing CRN. Unlike in underlay CRN where the interference limit is set, PUs and SUs actively seek opportunities to collaborate with each other. In [12], PD-NOMA is employed in large-scale CR network with randomly deployed PUs. New closed-form expressions of the outage probability of the NOMA users are derived to evaluate the performance of the considered CR-NOMA network. The application of NOMA to multicast cognitive radio networks (termed as MCR-NOMA) is investigated in [13]. Here, a dynamic cooperative MCR-NOMA scheme in which the multicast secondary users serve as relays to improve the performance of both primary and secondary networks is proposed. Based on the available channel state information (CSI), three different secondary user scheduling strategies for the cooperative MCR-NOMA scheme are presented. Results demonstrate significant performance gains for both networks.

9.1.2 Motivation of the Proposed Model

The multiplexing of PUs and SUs on a cognitive PD-SCMA system comes with their own challenges: PU vs SU power allocation (PA), PUs vs SUs codebook assignment (CA), PUs vs SUs pairing interference among others. These challenges have not been addressed by the above-mentioned works. For the system to be effective, proper resource (i.e., codebooks and power) allocation and user pairing (UP) schemes are required to keep the multiple interferences at the required levels to maintain acceptable quality of service (QoS) at decent receiver complexity. Only then we can realize the benefits of applying hybrid PD-SCMA scheme on a CRN. Analytical evaluation of the complex system featuring CRN with both PUs and SUs employing multiple QoS schemes (CA, PA, UP) is not trivial and cannot be easily evaluated by the exhaustive search methods.

9.1.3 Contribution of the Work

This work seeks to integrate PD-SCMA NOMA capabilities into CR concepts to achieve controllable interference for more intelligent spectrum sharing in underlay cognitive radio. The idea is to superimpose a PU and a SU into a codebook, while allocating different power levels and hence apply SIC decoding at the receiver. Furthermore, general PD-NOMA features power allocation to user on one resource unit channel gain sorting order strategy. In CR-PD-SCMA, power allocation is done considering all the resource units used by users for the sorting purpose, resulting in an optimal SIC ordering. Different from [10], this work proposes for sum-rate maximization through codebook sharing between SUs and PUs with minimal interference. RA for optimizing the achievable sum-rate for such a system is one of the main focus of this work. By decomposing the original sum-rate optimization problem into CA, UP and PA sub-problems, we propose dual parameter ranking resource allocation (DPR-RA) technique. DPR-RA utilizes the proposed ranking metrics for the dual players, i.e., users and resources, to rank and optimally match each other while employing a matching algorithm. We then propose an alternate search method (ASM) [14] at the transmitter to iteratively and jointly allocate individual resources to PUs and SUs. From the simulation results, the proposed DPR-RA schemes can improve the performance of the CR-PD-SCMA network by approximately 20% whereas providing fair allocation of resources to users.

9.2 System Model and Problem Formulation

9.2.1 System Model

The network model of Fig. 9.1 is for an uplink hybrid PD-SCMA based underlay CRN with a centralized macro base station (MBS) serving M randomly distributed primary users (PUs). The macro cell is overlaid by F secondary networks, each

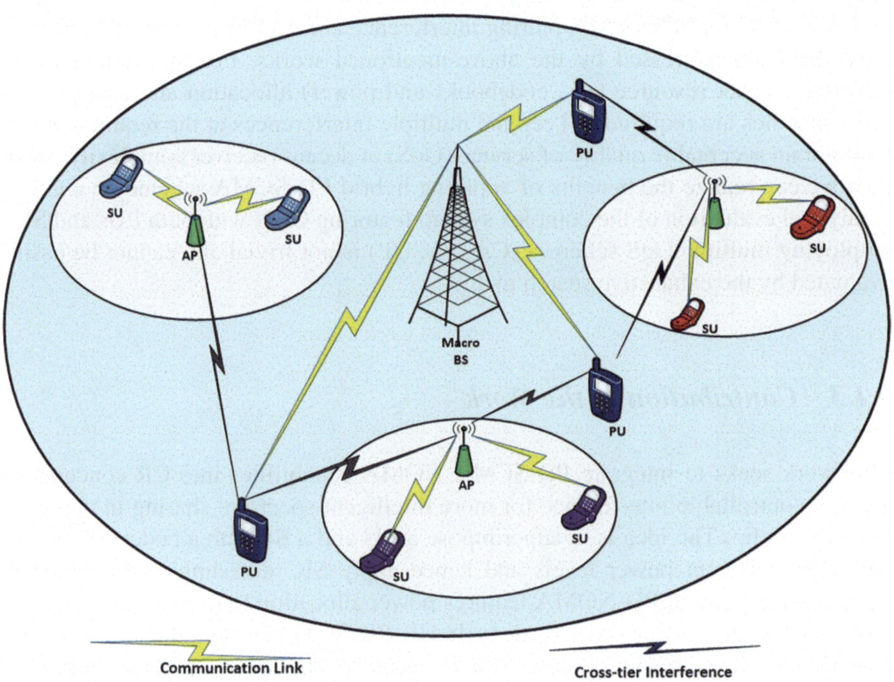

Fig. 9.1 System model of an uplink PD-SCMA underlay cognitive radio network

characterized by a centralized low power access points (APs) serving K uniformly distributed SUs. PUs and SUs are co-multiplexed over the same time-frequency N resource units (RUs) for enhanced spectral efficiency. Particularly, the PUs, SUs and APs are equipped with a single antenna for simplicity in analysis. The total network bandwidth B is equally shared by the N RUs. We define the following policies:

- The PA policy $\mathbf{P} = \left\{P_{F,K,C}^{SU}, P_{M,C}^{PU}\right\}$ is such that the transmitter allocates $P_{F,K,C}^{SU} = \left[P_{f,k,c}^{SU}\right]_{F\times K\times C}$ power to the kth SU on codebook c of the fth small cell and $P_{M,C}^{PU} = \left[P_{m,c}^{PU}\right]_{M\times C}$ to the mth PU utilizing the same codebook c. Note that $P_{m,c}^{PU} >> P_{f,k,c}^{SU}$.
- The CA policy $\mathbf{Q} = \left\{Q_{M,C}^{PU}, Q_{F,K,C}^{SU}\right\}$ where $Q_{M,C}^{PU} = \left[q_{m,c}^{PU}\right]_{M\times C}$ and $Q_{F,K,C}^{SU} = \left[q_{f,k,c}^{SU}\right]_{F\times K\times C}$ denote the macro cell and AP transmitter codebook assignment matrix, respectively. Also, $\left[q_{m,c}^{PU}\right] = 1$ implies the mth PU is assigned codebook c, otherwise $\left[q_{m,c}^{PU}\right] = 0$. Similarly, $\left[q_{f,k,c}^{SU}\right] = 1$ means that codebook c is assigned to the kth SU in the fth AP.
- The UP policy $\mathbf{A} = \left[A_{k,m}^{c}\right]_{K\times M}$, where $A_{k,m}^{PU} = 1$ denotes that the kth SU is paired with the mth PU on codebook c, while $A_{k,m}^{PU} = 0$ denotes otherwise. The policy matrix $\mathbf{A}$ pairs $Q_{M,C}^{PU}$ with $Q_{F,K,C}^{SU}$.

In a hybrid PD-SCMA underlay cognitive model, resource allocation (RA) plays an essential role in the signal recovery and throughput performance at the receiver. A codebook consists of d_v RUs and therefore, the proposed RA strategy must account for the effects of all RUs in a codebook. This work adopts the root mean square (RMS) of d_v RUs to determine the codebook channel gain for the mth PU and kth SU of fth secondary network on codebook c and is given by $h_{m,c}^{PU} = \sqrt{\frac{1}{d_v}\sum_{n\in N_c}\left|h_{m,n}^{PU}\right|^2}$ and $h_{f,k,c}^{SU} = \sqrt{\frac{1}{d_v}\sum_{n\in N_c}\left|h_{f,k,n}^{SU}\right|^2}$ where $h_{m,c}^{PU}$ and $h_{f,k,c}^{SU}$ denote channel gains from the mth PU and the kth SU on the nth RU link in the fth secondary network, respectively. A hybrid PD-SCMA multiplexing model for the SUs and PUs is with$N = 4$, $d_v = 2$ as illustrated in Fig. 9.2.

The co-tier interference exhibited in each secondary network is due to the SCMA resource structure. However, cross-tier interference is exhibited over codebooks in which PUs are paired with SUs. At the transmitter, users are paired based on the channel quality of the codebooks. At the receiver, joint SIC-MPA based receiver is employed to recover the codebooks of superimposed users. Firstly, MPA is deployed to decode the signals transmitted via each codebook. Secondly, SIC is deployed for separating user signals transmitted via the same codebook. For optimal SIC decoding at the AP, $P_{m,c}^{PU}\left|h_{m,c}^{PU}\right|^2 > P_{f,k,c}^{SU}\left|h_{f,k,c}^{SU}\right|^2$.

Prior to detection of the kth SU signal on codebook c, the AP first decodes the mth PU signal superimposed on that codebook and then removes $m-1$ poor channel PUs from the observation $y_{f,k,c}$ while treating PUs with better channel conditions as interference. Consequently, after detecting the signals of all PUs with worse channel

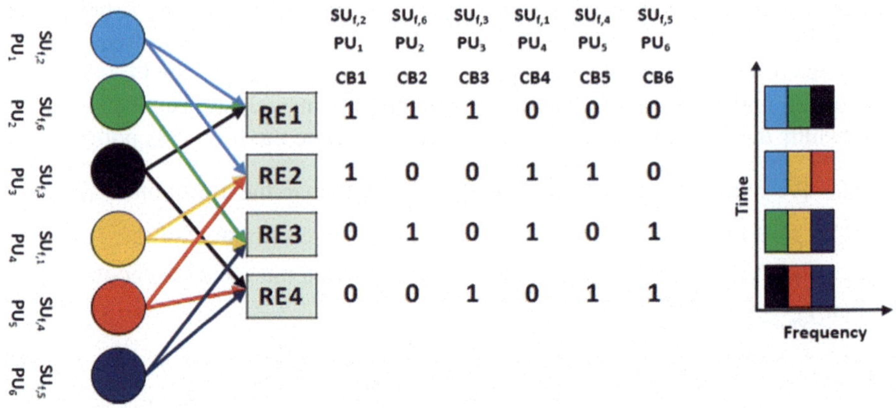

Fig. 9.2 Multiplexing model for RU mapping, SU codebook assignment and PU pairing with $N = 4, d_v = 2$

qualities than the corresponding SU, the available signal at the fth AP from the kth SU on codebook c is given by

$$y_{f,k,c} = q^{SU}_{f,k,c} h^{SU}_{f,k,c} \sqrt{P^{SU}_{f,k,c}} x_k + \sum_{m=1}^{M} A^{PU}_{m,c} h^{PU}_{m,c} \sqrt{P^{PU}_{m,c}} x_m + \sigma^2_{f,k,c}, \tag{9.1}$$

where x_k and x_m are the transmitted signals by the kth SU and mth PU, respectively, $\sigma^2_{f,k,c}$ denotes the Additive Gaussian noise power (AWGN) of the kth SU in the fth secondary network on codebook c. The instantaneous signal to interference plus noise ratio (SINR) of the mth PU at the fth SBS, $\gamma^{PU}_{m,c}$ is given by

$$\gamma^{PU}_{m,c} = \frac{A^{PU}_{m,c} P^{PU}_{m,c} |h^{PU}_{m,c}|^2}{\sum_{j=m+1}^{J} A^{PU}_{j,c} P^{PU}_{j,c} |h^{PU}_{j,c}|^2 + q^{SU}_{f,k,c} P^{SU}_{f,k,c} |h^{SU}_{f,k,c}|^2 + \sigma^2_{m,c}}, \tag{9.2}$$

while that of the kth SU $\gamma^{SU}_{f,k,c}$, after successful SIC of all PUs in each codebook is given by

$$\gamma^{SU}_{f,k,c} = \frac{q^{SU}_{f,k,c} P^{SU}_{f,k,c} |h^{SU}_{f,k,c}|^2}{\sum_{m=1}^{M} A^{PU}_{k,m} P^{PU}_{m,c} |h^{PU}_{m,c}|^2 + \sigma^2_{f,k,c}}. \tag{9.3}$$

The SUs achievable data rate, $R^{SU}_{f,k,c}$ is given as

$$R^{SU}_{f,k,c} = \log_2 \left(1 + \gamma^{SU}_{f,k,c}\right), \tag{9.4}$$

while that of the PUs is similarly derived and given as

$$R_{m,c}^{PU} = \log_2\left(1 + \gamma_{m,c}^{PU}\right), \tag{9.5}$$

9.2.2 *Problem Formulation*

Considering perfect channel state information, power and spectrum resource allocation is performed by AP for each SU. This work targets at maximizing the total sum-rate under the various QoS requirements in the secondary networks. Therefore, the corresponding optimization problem is expressed as

$$\begin{aligned}
\max_{\mathbf{Q},\mathbf{A},\mathbf{P}} \quad & \mathbf{R}_{tot}(\mathbf{Q},\mathbf{A},\mathbf{P}) = \max_{\mathbf{Q},\mathbf{A},\mathbf{P}} \sum_{k=1}^{K} R_{f,k,c}^{SU}(\mathbf{Q},\mathbf{A},\mathbf{P}) \\
\text{s.t.} \quad & \mathbf{C1}: \sum_{k=1}^{K} R_{f,k,c}^{SU} \geq R_{min}^{SU}, \ \forall f \\
& \mathbf{C2}: \sum_{m=1}^{M} A_{k,m}^{PU} P_{m,c}^{PU} \left|h_{f,k,c}^{SU}\right|^2 \leq \mathcal{I}_{m\to k}^{c,th}, \ \forall k \\
& \mathbf{C3}: P_{f,k,c}^{SU} \geq 0, \ \forall f,k,c \\
& \mathbf{C4}: P_{m,c}^{PU} \geq 0, \ \forall m \\
& \mathbf{C5}: A_{k,m}^{PU} P_{m,c}^{PU} \left|h_{m,c}^{PU}\right|^2 \geq q_{f,k,c}^{SU} P_{f,k,c}^{SU} \left|h_{f,k,c}^{SU}\right|^2, \ \forall f,k,c,
\end{aligned} \tag{9.6}$$

where constraint **C1** sets the QoS rate requirement for SU; **C2** sets the tolerable pairing interference bound over each codebook; **C3** and **C4** ensures that the SU and PU transmit power are non-negative and finally **C5** guarantees successful interference cancellation at kth SU. The above problem is mixed non-convex and combinatorial in nature and therefore prohibitive. To solve (9.6), a DPR-RA method is proposed and employed as in the subsequent section.

9.3 Dual Parameter Ranking Resource Allocation (DPR-RA)

The objective function (9.6) is decomposed into three sub-problems, CA, UP and PA. The individual resource allocation policies are formulated as dual parameter ranking (DPR) problems and solved using a one-to-many matching game algorithm proposed for stable matching [15]. The DPR observes players X and Y of sets $\mathcal{X}$ and $\mathcal{Y}$ respectively, ranks and matches them accordingly based on the preference

Algorithm 1: Transmitter ASM-RA Algorithm

Initialization: Initialize the matrix $\mathbf{Q}(0)$, $\mathbf{A}(0)$ and $\mathbf{P}(0)$ at iteration $t = 0$ and maximum iterations t_{max};

```
while t ≤ t_max or Convergence is false do
    Compute and allocate power P(t) to PUs;
    while APs f = 1 : F do
        Compute and allocate power P(t) to SUs;
        Compute CA policy Q(t) for SU-codebook assignment;
    end
    Compute UP policy A(t) for SU-PUs pairing ;
    Set t = t + 1 and update Q(t), A(t), P(t).;
end
```

metrics. Let the metrics $\Theta_X(Y) = \psi^{RA}_{Y\to X}$ denote the ranking of player Y by player X and $\Theta_Y(X) = \psi^{RA}_{X\to Y}$ denote the ranking of player X by player Y. The preference metrics $\Theta_X(Y)$ and $\Theta_Y(X)$ are defined for each RA policy, $RA = CA, UP, PA$ in the subsequent sub-sections.

A DPR-RA matching game-based algorithm for the RA policies is presented in Algorithm 2. The convergence of the DPR-RA algorithm can be verified by observing the preference formulation of players. Preference relations of players $X \in \mathscr{X}$ and $Y \in \mathscr{Y}$ are fixed for a given power allocation. Hence, given fixed preference relations of $\mathscr{X}$ and $\mathscr{Y}$, the DPR-RA algorithm is known as the deferred acceptance algorithm in the two-sided matching which converges to a stable matching μ^*_{RA} [16]. Given the DPR-RA policies and employing the alternate search method (ASM) [14], the transmitter iteratively and jointly allocates individual resources as in Algorithm 1.

9.3.1 *Codebook Assignment,* Q

The RUs to codebooks mapping matrix ρ are usually predetermined and fixed [2, 11]. In the DPR-CA problem, codebook set $\mathscr{C}$ represents set of players $\mathscr{X}$, while set of SUs $\mathscr{K}$ represents set of players $\mathscr{Y}$. A codebook set $c \in \mathscr{C}$ ranks the SUs employing the achievable rate $R^{SU}_{f,k,c}$ derived in (9.4) while SU $k \in \mathscr{K}$ ranks the codebooks employing a channel quality-based ranking metric (9.1) [8]. The ranking metrics are given by

$$\psi^{CA}_{c\to k} = R^{SU}_{f,k,c} \tag{9.7}$$

$$\psi^{SU}_{f,k,c} = \sqrt{\frac{1}{d_v}\sum_{i\in N_c}\frac{P^{SBS}_{max}}{K}\frac{|g^{SU}_{f,k,i}|^2}{\sigma^2_{f,k,i}}}, \tag{9.8}$$

where P_{max}^{SBS} denotes maximum SBS transmit power and N_c is the set of d_v RUs utilized by codebook c. Consequently, for the DPR-CA, $\Theta_X(Y) = \psi_{k\to c}^{CA}$ and $\Theta_Y(X) = \psi_{c\to k}^{CA}$. It can be deduced that the metric $\psi_{c\to k}^{CA}$ is maximized for a SU with high gain on most of the RUs and close to the BS and minimized for a poor gain SU at the cell edge. Similarly, a codebook prefers a SU with high achievable rate. The output of the algorithm is a stable codebook-SU matching.

9.3.2 *User Pairing,* A

The SU-PUs pairing is defined by the pairing policy matrix **A** such that $A = \left[\mathbf{A}_{m,c}^{PU} = 1\right]_{K\times M}$. Similar to DPR-CA, the DPR-UP problem is modelled as a one-to-many matching game with set $\mathscr{X}$ and set $\mathscr{Y}$ representing SU set $\mathscr{K}$ and PU set $\mathscr{M}$, respectively. The ranking metrics are given as follows:

$$\psi_{m\to k}^{UP} = R_{f,k,c}^{SU} \tag{9.9}$$

$$\psi_{k\to m}^{UP} = \frac{P_{m,c}^{PU}\left|h_{m,c}^{PU}\right|^2}{\Delta_m^c} - \beta\delta_k P_{m,c}^{PU}\left|h_{m,c}^{PU}\right|^2, \tag{9.10}$$

where the first term in (9.10) represents PU m channel gain with $\Delta_m^c = 2^{R_{m,c}^{min}} - 1$, the SINR threshold corresponding to the minimum PU rate requirement, $R_{m,c}^{min}$. The second term measures the relative loss that PU m advances to the SU k over the codebook c, defined as the interference cost imposed by SBS to PU m. β is a fixed coefficient with unit interference of MBS due to the SU on codebook c. The quantity δ_k is given by $\delta_k = \max\left(0, \left(\sum_{m=1}^{M} A_{k,m}^{PU} P_{m,c}^{PU}\left|h_{f,k,c}^{SU}\right|^2 - \mathscr{I}_{m\to k}^{c,th}\right)/\mathscr{I}_{m\to k}^{c,th}\right)$. It quantifies the degree of violation of (9.9) at the SBS. The PUs rank SUs based on SUs achievable rate-based metric $R_{f,k,c}^{SU}$ (9.4). Consequently, for the DPR-UP $\Theta_X(Y) = \psi_{k\to m}^{UP}$ and $\Theta_Y(X) = \psi_{m\to k}^{UP}$ while $J \geq 1$ represents the PU pairing quota. The matching game results in user pairing matrix $A = [A_{k,m}^{PU}]_{K\times M}$.

9.3.2.1 Power Allocation, P

The DPR-PA provides the PA policy P that allocates power to SUs and PUs. In DPR-PA, the set of power levels P represents the set of players X, while the user sets i.e., SUs, $\mathscr{K}$ and PUs, $\mathscr{M}$ individually represent the set of players Y. The ranking metrics are given as follows:

$$\psi_{p\to k}^{PA} = 2^{R_{f,k,c}^{SU}} - 1 \tag{9.11}$$

Algorithm 2: DPR-RA algorithm

Initialization: Initialize;
Sets $\mathscr{X}$, $\mathscr{Y}$ and **H**, J;
Set of matched Y players,$\mathsf{Y}_X = \emptyset$;
Set of requested Y players $\mathsf{Y}_X^{req} = \emptyset$;
Set of rejected Y players $\mathsf{Y}_X^{rej} = \emptyset$;
Initialize bid request $b_{Y \to X}^{RA}(t)$;
Evaluate stable matching μ_{RA}^* Each player $Y \in \mathscr{Y}$ constructs $\succ_{Y,RA}$ using the metric $\Theta_Y(X)$.;
while $\sum_{\forall X,Y} b_{Y \to X}^{UP}(t) \neq 0$ **do**
    for *each unassociated Y:* **do**
        Find $X = \arg\max_{X \in \succ_{Y,RA}} \Theta_Y(X)$;
        Send a request $b_{Y \to X}^{RA}(t) = 1$ to player X;
        for *each player X* **do**
            Find $X = \arg\max_{X \in \succ_{Y,RA}} \Theta_Y(X)$;
            Update $\mathsf{Y}_X^{req} \leftarrow \{Y : b_{Y \to X}^{RA}(t) = 1, Y \in \mathscr{Y}\}$;
            Construct $\succ_{Y,RA}$ using the metric $\Theta_X(Y)$
        end
        if $\mathsf{Y}_X \leq J$ **then**
            $\mathscr{Y}_X \leftarrow \mathsf{Y}_X$;
        else
            repeat ;
            Accept $Y = \arg\max_{Y \in \succ_{X,RA}} \sum_{Y \in \mathscr{Y}_X} \Theta_X(Y)$;
            Update $\mathscr{Y}_X \leftarrow \mathscr{Y}_X \cup Y$;
            Until $|\mathscr{Y}_X| = J$;
        end
        Update $\mathsf{Y}_X^{rej} \leftarrow \left\{\mathsf{Y}_X^{rej} \setminus \mathsf{Y}_X\right\}$;
        Remove player $X \in \succ_{Y,RA}, \forall Y \in \mathscr{Y}_X^{rej}$;
    end
end
Output: a stable matching μ_{RA}^*

$$\psi_{p \to m}^{PA} = 2^{R_{m,c}^{PU}} - 1 \tag{9.12}$$

$$\psi_{k \to p}^{PA} = \begin{cases} \min\left(\bar{P}_{f,k,c}^{SU}, \max\left(P_{f,k,c}^{SU}, P_{f,k,c}^{SU,min}\right), P_{f,k,c}^{max}\right), & \text{if } \Delta \geq P_{f,k,c}^{SU,min} \\ Infeasible, & \text{otherwise} \end{cases} \tag{9.13}$$

$$\psi_{m \to p}^{PA} = \min\left(P_{m,c}^{PU}, P_{max}^{PU}\right), \tag{9.14}$$

where in (9.11) and (9.12), $\psi^{PA}_{p \to i}$, $i \in \{m, k\}$ gives the achievable SINR-based metric employed by the power levels to rank users $i \in \{m, k\}$. The users rank the power levels based on a metric $\psi^{PA}_{i \to p}$, determined by the individual sets of PUs and SUs. In (9.13), $\psi^{PA}_{k \to p}$ represents the QoS aware power allocation metric [8] employed by the SUs to rank the power levels $P^{SU}_{f,k,c}$. In (9.14), $\psi^{PA}_{m \to p}$ represents the minimum SINR based metric employed by the PUs to rank the power levels $P^{PU}_{m,c}$, given by

$$P^{PU}_{m,c} = \frac{2^{\frac{R^m_{min}}{B_{ru}}} - 1}{q^{PU}_{m,c}\left|h^{PU}_{m,c}\right|^2}\left(q^{SU}_{f,k,c}P^{SU}_{f,k,c}\left|h^{SU}_{f,k,c}\right|^2 + \sigma^2_{m,c}\right), \tag{9.15}$$

where P^{PU}_{max} denotes the maximum PU power requirement. Consequently, for the DPR-PA $\Theta_X(Y) = \psi^{PA}_{p \to i}$ and $\Theta_Y(X) = \psi^{PA}_{i \to p}$, $i \in \{m, k\}$ and $J = 1$.

9.4 Results and Discussion

The sum-rate performance of the proposed DPR-RA based hybrid CR-PD-SCMA is evaluated in comparison with conventional RA schemes. They include biological RA-based PD-SCMA [9], sub-modularity and SCA-based RA scheme [14], efficient RA scheme of [15], and lastly the worst-case scenario RA of equal PA, random UP and regular CA. In the simulations, primary users are randomly distributed within the macro cell coverage area, while secondary users are uniformly distributed within the serving underlay AP coverage area. The set system parameters include macro cell radius $= 1$ km, secondary network radius $= 20$ m, $B = 10$ MHz, $\sigma^2_{f,k,c} = \frac{B}{N}N_0$ where $N_0 = -174$ dBm/Hz is the AWGN power spectral density, and interference threshold, $\mathscr{I}^{c,th}_{m \to k} = 10^{-5.5}$ W.

Figure 9.3 illustrates the overall system sum-rate performance for the proposed matching game-based DPR-RA algorithms, biological based RA algorithms in PD-SCMA [9], sub-modularity-based CA+UP with SCA-DC PA of [14], efficient resource management schemes of [15] and lastly equal power and regular CB assignment schemes versus the number of SUES. By varying the number of SUEs from 4 to 24, it can be observed that the system sum-rate increases as the number of SUEs increases in all the schemes. DPR-RA schemes outperform other schemes due to non-exclusive allocation of codebooks performed at the transmitter therefore eliminating the codebook interference. Additionally, the RA schemes result in better sum-rate performance than the equal power and regular CB assignment attributed to the efficient spectral RU utilization associated with the RA schemes.

Figure 9.4 depicts the CR-PD-SCMA system sum-rate performance based on the proposed DPR-RA schemes employed against signal to noise ratio (SNR). The results show that the sum-rate increases monotonically with the SNR. In fact, it can be observed that CR-PD-SCMA with combined DPR-RA schemes achieve a higher

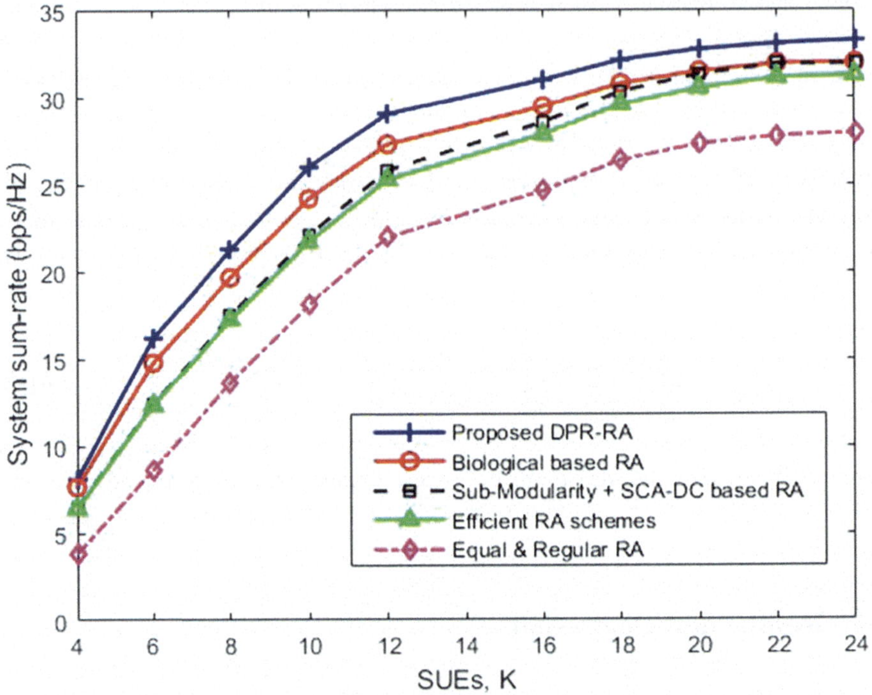

Fig. 9.3 Sum-rate vs number of SUEs for different schemes

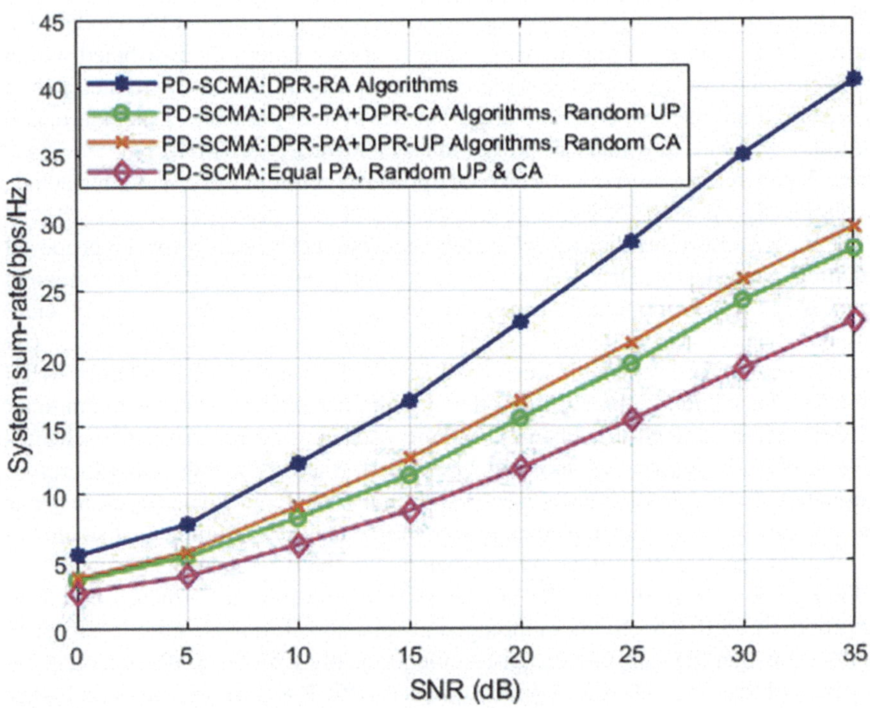

Fig. 9.4 Sum-rate vs SNR for CR-PD-SCMA RA schemes

sum-rate compared to the performance when any of the other schemes is used. This confirms the significant performance improvement of applying the DPR-CA and DPR-UP schemes, albeit higher performance with DPR-UP scheme, thereby ascertaining the importance of optimal pairing of users in a codebook which gives a considerate performance improvement as compared to the applied CA scheme.

To evaluate the fairness of the algorithms in allocating the resources among users in the network, Jain's fairness metric, J [17] is employed and defined as

$$J = \frac{\left(\sum_{k=1}^{K} R_{f,k,c}^{SU}\right)^2}{K \times \sum_{k=1}^{K} \left(R_{f,k,c}^{SU}\right)^2}. \tag{9.16}$$

From (9.16), the index has a range of $\frac{1}{K}$ (no fairness) to 1 (perfect fairness). In Fig. 9.5, the fairness performance of the considered algorithms is outlined. The DPR-RA is observed to outperform other schemes in terms of fairness as it has higher fairness index overall. This implies that the DPR-RA matches resources to users fairly. Its performance is followed by biological RA based CR-PD-SCMA and sub-modularity plus SCA-DC based RA with the worst-case scenario RA i.e., equal PA with random UP and regular CA showing the worst performance.

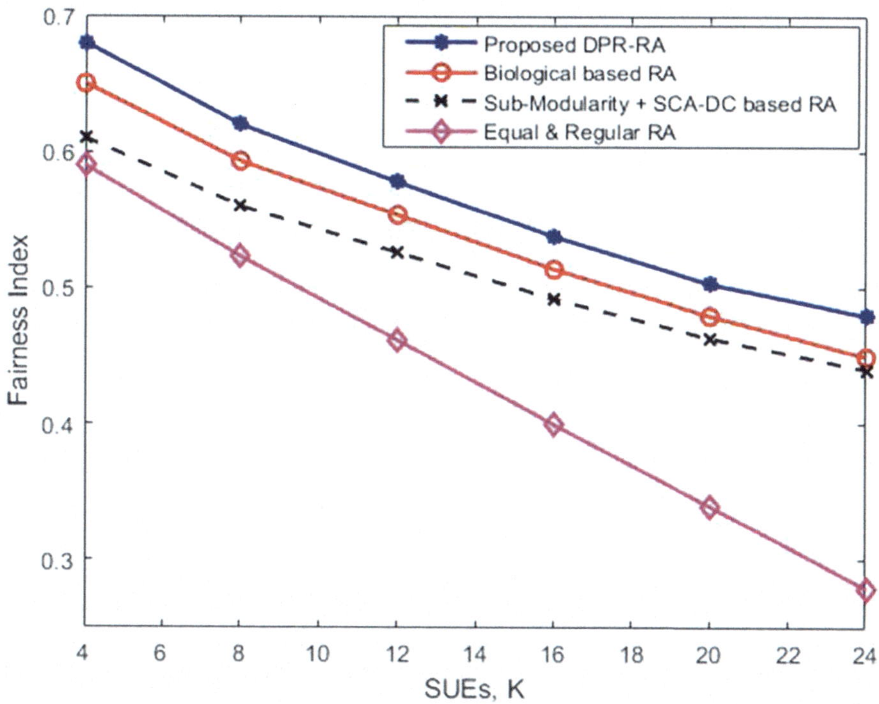

Fig. 9.5 Fairness vs number of users

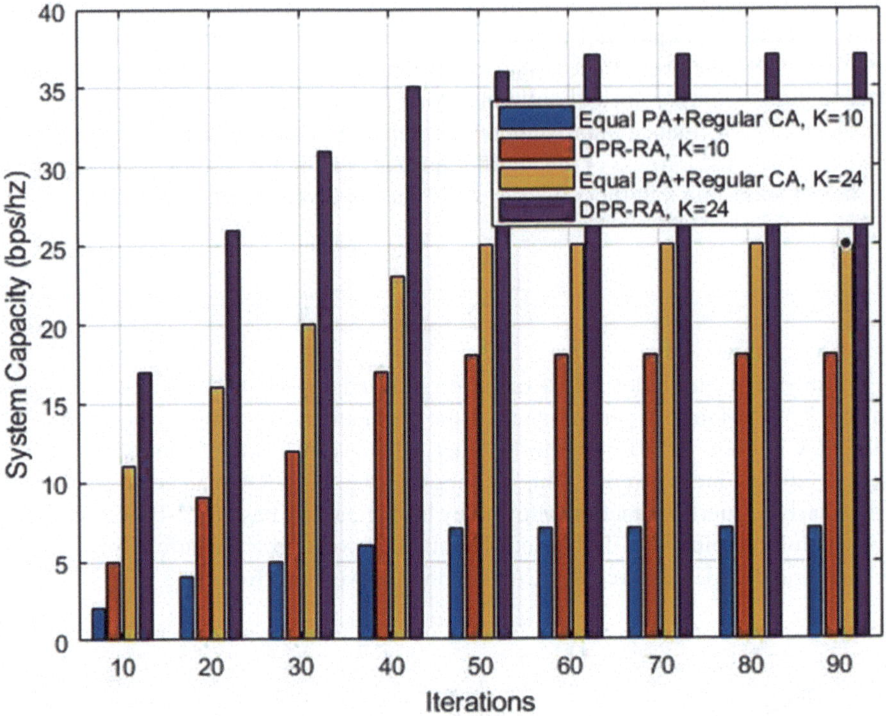

Fig. 9.6 Convergence of the proposed ASM algorithm

Figure 9.6 shows the achievable sum-rate value versus the number of iterations for different number of SUEs with only one MUE multiplexed in each codebook. From the figure, it can be deduced that after few iterations, the achievable sum-rate value converges, reiterating the convergence of the ASM algorithm [14]. Nonetheless, CR-PD-SCMA employing DPR-RA schemes converge to an optimal solution with higher sum-rate compared to CR-PD-SCMA system utilizing equal PA, uniform UP and regular CA.

9.5 Conclusions

In this work, resource allocation for uplink hybrid PD-SCMA system in CR for sum-rate performance is investigated. Due to the prohibitive mixed non-convex combinatorial nature of the maximization problem, a dual parameter ranking based resource allocation employing one to many matching algorithm is proposed which results in a stable matching between users and resources. At the transmitter, the individual resources (codebooks, pairs and power) are iteratively allocated to users using the alternative search method which is proven to converge within few iterations. From simulation results, DPR-RA based hybrid CR-PD-SCMA demonstrates significant sum-rate performance improvement in comparison with conventional RA schemes and even better performance compared to the worst-case scenario of equal PA, random UP and regular CA. Better sum-rate performance can certainly be achieved with optimized SIC operation. Further work will be to investigate the performance of succeeding models featuring more multiplexed PUs on a codebook and extended aspects like signalling overhead, channel uncertainty and many others for conclusive deductions.

References

1. Yuanwei, L., Zhijin, Q., et al. (2017). Non-orthogonal multiple access in large-scale heterogeneous networks. *IEEE Journal on Selected Areas in Communications, 35*(12), 2667–2680.
2. Almalfouh S., & Stüber, G. (2011). Interference-aware radio resource allocation in OFDMA-based cognitive radio networks. *IEEE Transactions on Vehicular Technology, 4*, 1699–1713.
3. Lv, L., & Chen, J., et al. (2018). Cognitive non-orthogonal multiple access with cooperative relaying: a new wireless frontier for 5G spectrum sharing. *IEEE Communications Magazine, 56*(4), 188–195.
4. Ali, M. S., Tabassum, H., Hossain, E. (2016). Dynamic user clustering and power allocation for uplink and downlink non-orthogonal multiple access (NOMA) systems. *IEEE Access, 4*, 6325–6343.
5. Xin, S., Aniello, C., Christian, E., & Chang, C. (2018). Power domain NOMA to support group communication in public safety networks. *IEEE Access, 84*, 228–238.
6. Alnoman, A., Erkucuk, S., & Anpalagan, A. (2019). Sparse code multiple access-based edge computing for IoT systems. *IEEE Internet of Things Journal, 1*, 1–2.
7. Balasubramanya, N., Payami, S., & Sellathurai, M. (2018). Uplink resource allocation for shared LTE and SCMA IoT systems. In *IEEE 87th Vehicular Technology Conference (VTC Spring), Porto* (pp. 1–5).
8. Chege, S., & Walingo, T. (2020). Energy efficient resource allocation for uplink hybrid power domain sparse code nonorthogonal multiple access heterogeneous networks with statistical channel estimation. *Transactions on Emerging Telecommunications Technologies, 32*, e4185.
9. Sefako, T., &Walingo, T. (2020). Biological resource allocation algorithms for heterogeneous uplink PD-SCMA NOMA networks. *IEEE Access, 8*, 194950–194963.
10. Moltafet, M., & Mokari, N., et al. (2018). A new multiple access technique for 5G: Power domain sparse code multiple access (PSMA). *IEEE Access, 6*, 747–759.
11. Lv, L., Ni, Q., & Chen, J. (2017). Application of non-orthogonal multiple access in cooperative spectrum-sharing networks over Nakagami- *m* fading channels. *IEEE Transactions on Vehicular Technology, 66*(6), 5506–5511.
12. Liu, Y., Ding, Z., Elkashlan, M., & Yuan, J. (2016). Nonorthogonal multiple access in large-scale underlay cognitive radio networks. *IEEE Transactions on Vehicular Technology, 65*(12), 10152–10157.
13. Lv, L., Ni, Q., Chen, J., & Ding, Z. (2017). Design of cooperative non-orthogonal multicast cognitive multiple access for 5G systems: User scheduling and performance analysis. *IEEE Transactions on Communications, 65*(6), 2641–2656.
14. Zakeri, A., Moltafet, M., & Mokari, N. (2019). Joint radio resource allocation and SIC ordering in NOMA-based networks using submodularity and matching theory. *IEEE Transactions on Vehicular Technology, 68*(10), 9761–9773.

15. Sultana, A., Woungang, I., & Anpalagan, A., et al. (2020). Efficient resource allocation in SCMA-enabled device-to-device communication for 5G networks. *IEEE Transactions on Vehicular Technology*, *69*(5), 5343–5354.
16. LeAnh, T., et al. (2017). Matching theory for distributed user association and resource allocation in cognitive femtocell networks. *IEEE Transactions on Vehicular Technology*, *66*(9), 8413–8428.
17. Evangelista, J., Sattar, Z., Kaddoum, G., & Chaaban, A. (2018). Fairness and sum-rate maximization via joint channel and power allocation in uplink SCMA networks. Preprint. arXiv: 1805,11722.

Index

A
Access, cybersecurity, 22
ACIDS framework
  application layer
    application workflow, 93
    DDoS attack, 92
    DoS attack, 92
    framework, 86
    SQL Injection, 93
  communication layer
    eavesdropping, 89
    framework, 86
    jamming, 90
    Man-In-The-Middle Attack, 89–90
    protocol violation, 90–91
  data layer
    data theft, 91
    default or test accounts, 92
    framework, 86
    identity theft, 91
    unauthorized access control, 92
  infrastructure layer
    device hijacking, 88
    framework, 85–86
    spying, 88–89
    theft, 88
  stakeholders
    framework, 86–87
    threats, 93–94
Active Jammers, 90
Adafruit VL53L0X time-of-flight distance sensor, 3
Additive Gaussian noise power (AWGN), 128
Aerial robots, 99
Alternate search method (ASM), 126, 130
Anomaly Detection loT (AD-IoT) system, 83
Application, Communication, Infrastructure, Data, and Stakeholders (ACIDS). *see* ACIDS framework
ARP Spoofing, 90
Artificial neural network, 50, 57, 58
Authentication, 21
Automatic patrol, 1
Availability, cybersecurity, 22
Average latency, 35

B
BIKE (code-based), 112
Binary class distribution, 47
Bitcoins, 80
Bluetooth, 100, 101, 104
Bluetooth Low Energy (BLE), 100–104, 106, 107, 109
Bluetooth wireless communication, 17
Borderline SMOTE, 48
Brno University Hospital's smart system, 80

C
Camera-based approaches, 100
Channel state information (CSI), 125
Check-and-Spray mechanism, 63, 65
Classic McEliece cryptosystem, 111–113
Closed-Circuit TV (CCTV), 32
Cloud services, 79
CloudSim, 32, 40
Codebook assignment, 130–131

I. Woungang, S. K. Dhurandher, *4th International Conference on Wireless, Intelligent and Distributed Environment for Communication*, Lecture Notes on Data Engineering and Communications Technologies 94, https://doi.org/10.1007/978-3-030-89776-5

Code division multiple access (CDMA), 123
Cognitive radio (CR)
convergence of proposed ASM algorithm, 136
CR-NOMA network, 125
CR-PD-SCMA, 125, 126, 133–137
DPR-RA (*see* Dual parameter ranking resource allocation (DPR-RA))
dynamic spectrum access, 123, 124
fairness *vs* number of users, 135
NOMA, 123–125
objective of, 124
PD-NOMA, 124, 125
PD-SCMA, 124–127, 133
problem formulation, 129
PUs *vs* SUs, 125
spectrum sending, 123, 124
sum-rate *vs* number of SUEs for different schemes, 134
sum-rate *vs* SNR for CR-PD-SCMA RA schemes, 134
system model, 126–129
Colorado Department of Transportation of Atlanta (CODT), 80
Confidentiality, 21
Convolutional neural networks, adaptive system
design methodology, 3–5
hardware and software implementation
data preprocessing, 6–7
electrical current sensing and adaptive control, 11
integrated development environment, 5–6
mobile subsystem, 11–13
MoveIt Robot control steps, 10–11
object-picking subsystem, 8–10
remote monitoring, 7–8
image recognition, 13–14
APP user interface optimization, 15–16
confusion matrix, 14
ROS node optimization, 14–15
self-adaptive jaw damage-proof mechanism, 16–17
motivation, 2
system architecture, 1, 2
COVID-19 pandemic, 80
CRYSTALS-DILITHIUM, 111
CRYSTALS-KYBER, 111
Cyberattack, 21–23, 80
Cybercriminals, 80, 92
Cybersecurity attacks, 79, 83, 89, 93, 94
Cybersecurity data science, 21–22
algorithms for, 25–26
applications for, 26–28
concepts of, 22–23
subjects, 24
Venn diagram, 23–24
mechanisms, 22
supervised anomaly detection schemes, 28

D
Danger zone, data science, 24
DARPA, 27–28
Database, 1, 8
Data hacking skills, data science, 23
Data mining, 103, 106, 107, 109
Data science, 23–24
Data theft, 91
Decryption, 90
Default or test accounts, 92
Deferred acceptance algorithm, 130
Denial of Service (DoS) attack, 27, 92
Denning, 28
Detection rate (DR), 28
Device hijacking, 88
Digital Signature Algorithms, 111, 112
Dijkstra algorithm, 67
Distance sensor, 3–5
Distributed Denial of Service (DDoS) attack, 26–27, 91, 92
DJI Mavic drones, 103
DNS Spoofing, 90
Domain knowledge, data science, 23
Drone assist disaster management system, 102
Drones
accessibility of points, 107, 108
accuracy measurement, 109
advantages, 99
cluster 3D map points using K-means clustering algorithm, 106
drone assist disaster management system, 102
drone trajectory data mining, 106–107, 109
K-means clustering, 107, 108
project configuration
control center, 103–104
drone structure, 103
industrial building structure, 103
rescue operation outcome, 109
3D map using 2D map of victim's position (local map)
fuzzy logic, 105–106
image analysis, 105

mathematical equation and rules for reachable points, 104–105
2D map of victim's position (global map), 104
Drone trajectory data mining, 106–107, 109
Dual parameter ranking resource allocation (DPR-RA), 124, 126
algorithm, 130
codebook assignment, 130–131
objective function, 129
power allocation, 131–133
user pairing, 131

E
Eavesdropping, 89
Edge TPU Accelerator, 3, 4
EM-based clustering technique, 25
EmuFog, 40
Energy consumption, 35

F
FALCON, 111
False-positive rate (FPR), 28
Fiat Chrysler Automobiles, 80
First Come First Serve (FCFS), 34
Fog-based Traffic Surveillance Application model, 33
Fog computing application model topology, 36
FogNetSim++, 40
FogTorchII, 40
Fourth Industrial Revolution (4IR), 31
FrodoKEM (lattice-based), 112
Fuzzy-based Check-and-Spray Geocast Routing Protocol (FCSGRP)
notations, 64
real mobility traces model
average latency *vs.* buffer size, 74
average latency *vs.* TTL, 75
delivery ratio *vs.* buffer size, 73
delivery ratio *vs.* TTL, 74
overhead ratio *vs.* buffer size, 75
overhead ratio *vs.* TTL, 76
routing scheme, 65
SPMBM model
average latency ratio *vs.* number of hosts, 69
average latency *vs.* buffer size, 70
average latency *vs.* TTL, 71
delivery ratio *vs.* buffer size, 70
delivery ratio *vs.* learning rate, 72
delivery ratio *vs.* number of hosts, 69
delivery ratio *vs.* TTL, 71
overhead ratio *vs.* number of hosts, 72
Fuzzy Inference System (FIS), 101
Fuzzy logic for 3D mapping, 105–106

G
Gaussian elimination, 114, 115
Gaussian Mixture Model (GMM), 25
Gaussian reductions, 113, 114, 116–118, 120, 121
GeMSS (multivariate), 112
Generalized Priority (GP), 34
gf_mul_13 multiplication, 114
gf_mul_13_128's optimization, 117
Goppa code, 114
Graphical user interface (GUI), 40

H
Hague Security Delta, 94
HC-SR04 ltrasonic distance sensor, 3
Heterogeneous multi-tier network (HetNet), 124
HQC (code-based), 112
HTTPS Spoofing, 90

I
Identity theft, 91
iFogSim, 32, 40
Image analysis, 105
Industrial building structure, 103
INFOCOM 2006 dataset, 67
Information and communication technologies (ICT), 79, 86, 91
Information security, 22–23
Infrared sensor-based camera, 100, 101
Interception, 89–90
Internet Control Message Protocol (ICMP), 91
Internet of Things (IoT), 2, 31, 79
Internet Protocol (IP), 91
IP Spoofing, 89
IR depth camera, 100, 103, 104
Italian National Fire Corps, 99

J
Jain's fairness metric, 135
Jamming, 90
Jeep Cherokee, 80

K
Key-establishment algorithms, 111, 112
Key-pair generation phase, 112, 113, 119–121
K-means clustering, 101, 102, 106–108
k-nearest neighbours algorithms, 50, 57, 58

L
La Conchita mudslide, 99
Light Detection and Ranging (Lidar), 100–104
Lightweight messaging protocol, 2
Linear map application, 119
Linear map creation, 119
Log-domain message passing algorithm (log-MPA), 124
"London City Challenge," 94
London Digital Security Centre, 94
Long-term evolution (LTE) users, 124

M
MAC address, 90
Machine-assisted disaster management practices, 99
Machine learning, 3, 24
Macro base station (MBS), 126
Macro user equipment's (MUEs), 124
Mamdani Fuzzy Inference System, 101
Man-In-The-Middle (MITM) attack, 89–90
Math knowledge, data science, 23
mat_t'rows, 116
Mavic drones, 103
McEliece cryptosystem
  classic McEliece cryptosystem, 112–113
  memory consumption, 120–121
  private key, 114
    Gaussian reduction, 117–118
    polynomial multiplication, 117
  public key, 115–116
    Gaussian reduction, 118–119
    linear map application, 119
    linear map creation, 119
  software optimizations, 113
  specifications of XPS machine, 116–117
  timings of sequential tests, 119, 120
mceliece8192128 parameters, 114, 115, 121
Memory consumption, 120–121
Message Forwarding Set (MFS), 63
Message integrity, 21
Mirandola earthquake, 99
Mobile device applications, 7–8
MobilenetV1, 3
Mobile phone detection technologies, 100
Modified safe-level SMOTE algorithm, 49, 50, 53, 56–58
Modified SMOTE method, 48–50, 53, 54, 57, 58
MoveIt Robot, 10–11
M4RIE library, 114, 120, 121
Multi-Agent Reinforcement Learning Congestion Control (MARL-CC)
  Q-table, 66
  real mobility traces model
    average latency *vs.* buffer size, 74
    average latency *vs.* TTL, 75
    delivery ratio *vs.* buffer size, 73
    delivery ratio *vs.* TTL, 74
    overhead ratio *vs.* buffer size, 75
    overhead ratio *vs.* TTL, 76
  routing scheme, 67
  SPMBM model
    average latency ratio *vs.* number of hosts, 69
    average latency *vs.* buffer size, 70
    average latency *vs.* TTL, 71
    delivery ratio *vs.* buffer size, 70
    delivery ratio *vs.* learning rate, 72
    delivery ratio *vs.* number of hosts, 69
    delivery ratio *vs.* TTL, 71
    overhead ratio *vs.* number of hosts, 72
Multiple-input multiple-output (MIMO), 124

N
Naïve Bayesian algorithms, 50, 57, 58
National Institute of Standards and Technology (NIST), 111–113
Network security, 22–23
Network usage, 35
Non-orthogonal multiple access (NOMA), 123–125
Novice users, 94

O
Object-picking subsystem, 8–10
one_row, 116
OpenCV, 7
OpenMP, 115
Opportunistic networks (OppNets)
  FCSGRP (*see* Fuzzy-based Check-and-Spray Geocast Routing Protocol (FCSGRP))
  MARL-CC (*see* Multi-Agent Reinforcement Learning Congestion Control (MARL-CC))
  RLFGRP (*see* Reinforcement Learning-based Fuzzy Geocast Routing protocol (RLFGRP))
  store-carry-and-forward mechanism, 61
ops_t'columns, 116
Orthogonal frequency division multiple access (OFDMA) subcarrier, 123

Othogonal multiple access (OMA) techniques, 123–125

P
Pairing policy matrix, 131
Pan–tilt–zoom (PTZ), 37
Picnic (zero-knowledge), 112
Polynomial multiplication, 112–114, 117, 118, 120, 121
Post-Quantum Cryptography, 111, 112
Power allocation, P, 131–133
Power consumption, 2
Power domain sparse code non-orthogonal multiple access (PD-SCMA) scheme, 124–127, 133
Power supply subsystem, 12–13
#pragma omp barrier, 115
Primary users (PUs), 123, 125–131, 133, 137
PrivySharing, 83
Protocol violation, 90–91
Public-key encryption, 111, 112

Q
Q-learning mechanism, 62
Quality of service (QoS), 125
Quantum computing theory, 111

R
Radio resource allocation (RRA), 123
Random Forest machine learning algorithm, 83
Ranking metrics, 130, 131
Raspberry Pi, 5, 6, 12, 14–15
Reactive Jammers, 90
Real mobility traces model
  FCSGRP
    average latency *vs.* buffer size, 74
    average latency *vs.* TTL, 75
    delivery ratio *vs.* buffer size, 73
    delivery ratio *vs.* TTL, 74
    overhead ratio *vs.* buffer size, 75
    overhead ratio *vs.* TTL, 76
  MARL-CC
    average latency *vs.* buffer size, 74
    average latency *vs.* TTL, 75
    delivery ratio *vs.* buffer size, 73
    delivery ratio *vs.* TTL, 74
    overhead ratio *vs.* buffer size, 75
    overhead ratio *vs.* TTL, 76
  RLFGRP
    average latency *vs.* buffer size, 74
    average latency *vs.* TTL, 75
    delivery ratio *vs.* buffer size, 73
    delivery ratio *vs.* TTL, 74
    overhead ratio *vs.* buffer size, 75
    overhead ratio *vs.* TTL, 76
Reinforcement Learning-based Fuzzy Geocast Routing protocol (RLFGRP), 61
  notations, 62
  Q-value, 62, 63
  real mobility traces model
    average latency *vs.* buffer size, 74
    average latency *vs.* TTL, 75
    delivery ratio *vs.* buffer size, 73
    delivery ratio *vs.* TTL, 74
    overhead ratio *vs.* buffer size, 75
    overhead ratio *vs.* TTL, 76
  routing scheme, 63
  SPMBM model
    average latency ratio *vs.* number of hosts, 69
    average latency *vs.* buffer size, 70
    average latency *vs.* TTL, 71
    delivery ratio *vs.* buffer size, 70
    delivery ratio *vs.* learning rate, 72
    delivery ratio *vs.* number of hosts, 69
    delivery ratio *vs.* TTL, 71
    overhead ratio *vs.* number of hosts, 72
Remote monitoring, 7–8
Robotic assistants, 99
Robot Operating System (ROS) development, 6, 14–15
Round Robin (RR), 34

S
SABER, 111
Safe-Level SMOTE
  algorithm, 55
  confusion matrix evaluation technique, 57, 58
  performance comparison, 57
  research objectives, 49–50
  research question, 49
SamSam virus, 80
Search and rescue, 100, 101, 106, 109
Secondary users (SUs), 123, 125–128, 130, 131, 133

Sequence diagram, traffic surveillance application, 37–38
Servo motors, 11
Shortest Path Map-Based Movement (SPMBM), 67
  FCSGRP
    average latency ratio *vs.* number of hosts, 69
    average latency *vs.* buffer size, 70
    average latency *vs.* TTL, 71
    delivery ratio *vs.* buffer size, 70
    delivery ratio *vs.* learning rate, 72
    delivery ratio *vs.* number of hosts, 69
    delivery ratio *vs.* TTL, 71
    overhead ratio *vs.* number of hosts, 72
  MARL-CC
    average latency ratio *vs.* number of hosts, 69
    average latency *vs.* buffer size, 70
    average latency *vs.* TTL, 71
    delivery ratio *vs.* buffer size, 70
    delivery ratio *vs.* learning rate, 72
    delivery ratio *vs.* number of hosts, 69
    delivery ratio *vs.* TTL, 71
    overhead ratio *vs.* number of hosts, 72
  RLFGRP
    average latency ratio *vs.* number of hosts, 69
    average latency *vs.* buffer size, 70
    average latency *vs.* TTL, 71
    delivery ratio *vs.* buffer size, 70
    delivery ratio *vs.* learning rate, 72
    delivery ratio *vs.* number of hosts, 69
    delivery ratio *vs.* TTL, 71
    overhead ratio *vs.* number of hosts, 72
Short Job First (SJF), 34
Signal to interference plus noise ratio (SINR), 128
Signal to noise ratio (SNR), 133
SIKE (supersingular curves isogeny), 112
Simultaneous localization and mapping (SLAM) technology, 11, 18, 100
Singapore, cybersecurity, 93
Small cell user equipment's (SUEs), 124
Smart city framework
  ACIDS (*see* ACIDS framework)
  AD-IoT, 83
  popular attacks, 81–82
  taxonomy of research papers in security of, 84
Smart Grid, 94
Smart Transportation, 94
Smart Water and Waste Management, 94
SMOTE
  attribute loop of, 50, 51
  demonstration, 52
  function, 52
  modified function of attribute loop of, 52, 53
  modified SMOTE demonstration, 52, 53
Smurf Attack, 91
SPHINCS+ (hash-based), 112
Spying, 88–89
SQL Injection (SQLi), 93
SSL hijacking, 90
SSL stripping, 90
Stakeholders, 86–87, 93–94
Starwood Preferred Guest (SPG) database, 80
Statistics knowledge, data science, 23
Store-carry-and-forward mechanism, 61
Substantive expertise, data science, 23
Successive interference cancellation (SIC), 124, 125, 127, 128, 137
Supervised anomaly detection schemes, 28
Support Vector Machine, 50, 57, 58

T

TCP SYN Flood Attack, 91
Theft, 88
Time division multiple access (TDMA), 123
TLS authentication, 90
Tohoku earthquake, 99
Traditional research, data science, 24
Traffic surveillance application
  activity diagram, 38–39
  algorithms, 33–34
    First Come First Serve, 34
    Generalized Priority, 34
    motivation for, 34–35
    Round Robin, 34
    Short Job First, 34
  design, 36–37
  evaluation metrics, 35
  fog computing application model topology, 36
  model, 36–37
  performance evaluation
    configuration setup, 40–41
    iFogSim, 40
  policies, 32
  real-time fog applications, 32
  sequence diagram, 37–38

simulation
average loop delay, 41
energy consumption, 41–42
execution time, 42
network usage, 42–43
task offloading, 33
2/3/4G cellular networks, 100

U
UAV-assisted natural disaster management, 100
Ubuntu, 6
UML activity diagram, 38–39
Unauthorized access control, 92
Unmanned aerial vehicles. *see* Drones
User pairing (UP), 125, 131
US hospital management system, 80

V
Vision-based techniques, 100

W
Weak passwords, 94
WiFi, 17, 67, 86, 89, 100
Wireless communication, 2

Z
Zero-day attacks, 25

The manufacturer's authorised representative in the EU is Springer Nature Customer Service Centre GmbH, Europaplatz 3, 69115 Heidelberg, Germany. If you have any concerns regarding our products, please contact ProductSafety@springernature.com

Printed and bound by CPI Group (UK) Ltd, Croydon, CR0 4YY
01/10/2026
02226294-0002